Optimization and Intelligent Selection of Aluminum Hybrid Composites

By
Jinu Jani

TABLE OF CONTENTS

CHAPTER 1	INTRODUCTION	Page No.
1.1	Background	1
1.2	Modes of gear failures	2
1.2.1	Wear	2
1.2.2	Tooth breakage	3
1.2.3	Fatigue	4
1.2.4	Pitching	5
1.2.5	Scoring	5
1.2.6	Plastic flow	6
1.2.7	Thermal cycle fatigue	7
1.3	Stresses in gears	8
1.3.1	Contact stresses	9
1.3.2	Bending stresses	10
1.3.3	Torsional Stresses	10
1.4	Selection criteria for gear materials	11
1.4.1	Accuracy	11
1.4.2	Mating gear material	11
1.4.3	Chemical environment	12
1.4.4	Service temperature	12
1.4.5	Strength	13
1.4.6	Coefficient of friction	13
1.5	Gear materials	13
1.5.1	Metallic gears	13
1.5.2	Non-metallic gears	15
1.6	Composites as a gear material	16
1.6.1	Aluminum as a gear material	17
1.6.2	Advantages of aluminum gears	19
1.6.3	Pitfalls of aluminum gears	20

	1.7	Aluminium metal matrix nanocomposites	21
	1.8	Importance and motivation of the study	23
	1.9	Problem statement	24
	1.10	Organization of thesis	25
		References	26
		Publications	26
CHAPTER-2		**LITERATURE REVIEW**	
	2.1	Introduction	27
	2.2	Different methods for the fabrication of composites	28
	2.3	Literature on the production of composites	35
	2.4	Contact stress analysis	41
	2.5	Literature studies on contact stresses	44
	2.6	Relevant literature on the present study	47
	2.7	Summary of the literature	53
	2.8	Literature gaps	53
	2.9	Objectives of Proposed Research	54
CHAPTER-3		**ARCHITECTURE FOR IDENTIFICATION OF THE BEST MIXTURE**	
	3.1	Materials and coating	56
	3.1.1	Application of aluminum hybrid materials	56
	3.1.2	Silicon carbide coating	58
	3.1.3	Manufacturing information	59
	3.2	Proposed SRNMS for mix selection	60
	3.2.1	Process of SRNMS	62
	3.3	Methodology	67
	3.3.1	Fabrication of composites	67
	3.3.2	Microstructural investigation	70
	3.3.3	Tensile testing	71
	3.3.4	Flexural testing	71
	3.3.5	Hardness test	71
	3.3.6	Impact testing	71
	3.3.7	Wear test	72
	3.4.	Results and discussion	72
	3.4.1	Optimization	72
	3.4.2	Microstructural behavior of the materials	76
	3.4.3	Wear rate with varying load, sliding distance, and sliding Speed	79
	3.4.4	The friction coefficient and abrasive friction coefficient (AFC)	84
	3.4.5	Root bending stress	88
	3.4.6	Thermal expansion	89
	3.4.7	Surface roughness	90
	3.4.8	Comparison assessment	91

3.4.9	Discussion	93
3.5	Conclusion	97
CHAPTER-4	**PERFORMANCE EVALUATION OF COMPOSITE GEARS**	
4.1	Fabrication of spur gears	99
4.2	Test rig	101
4.3	Calculation of stress	105
4.4	3D solid modelling	107
4.5	Finite element analysis	109
4.6	Result and discussion	110
4.6.1	Contact strain	110
4.6.2	ANSYS	115
4.6.3	Summary	118
CHAPTER-5	**CONCLUSIONS AND FUTURE SCOPE**	
5	Conclusions	119
5.1	Future scope	120

CHAPTER 1

1. INTRODUCTION

1.1. Background

Gears are fundamental mechanical components used to drive human ingenuity. Essentially, toothed wheels mesh to transfer motion or alter speed and direction rotationally. Figure 1.1 shows the transmission of motion through different sizes of gears. Their operation is simple: as one gear turns, it imparts motion to another, facilitating energy and motion transfer. This concept underlies gadgets ranging from ancient water mills to modern cars, clocks, and industrial machinery. Gears' efficiency depends on their design and arrangement, with size and tooth count deciding speed ratios. When a larger gear drives a smaller one (a pinion), the larger gear rotates slower but with greater force, while the smaller gear rotates faster but with less force. Engineers exploit this to tailor gear systems for speed, torque, or both, suiting diverse applications. Moreover, gears transcend mere force and speed transmission, enabling intricate setups like differentials and planetary systems. Planetary gears, comprising a central "sun" gear encircled by "planet" gears within an outer "ring" gear, find use in applications demanding precise speed and torque control, such as vehicle transmissions and industrial machinery. Differential gears, crucial for vehicles, allow wheels on the same axle to rotate at varying speeds during turns, ensuring smooth operation [1–7].

Gear failures pose operational and safety hazards, often stemming from excessive stress or poor mating. Heavy loads or uneven force distribution can compromise gear integrity, leading to surface damage or tooth failure. Misaligned gears, caused by manufacturing defects or improper installation, increase wear and fatigue, while friction between gear surfaces, exacerbated by inadequate lubrication, accelerates wear and heat buildup, risking gear functionality and life [8–11].

Figure 1.1 Mating gears

1.2. Modes of gear failures

Gear failures can be caused by various mechanisms and circumstances, which can have disastrous effects on gear systems and affect performance and efficiency. Comprehending the various gear failures is essential for pinpointing the root causes, putting preventative measures in place, and guaranteeing the dependability and durability of gear systems. The following is an explanation of a few standard modes of gear failures [12, 13, 22, 14–21]:

1.2.1. Wear

Wear, a significant gear failure mode, involves the gradual removal or displacement of material from the surface of gear teeth, influenced by various factors. Friction, lubrication, material properties, operating conditions, and environmental pollutants contribute to this process. Wear can manifest in different ways, including erosive wear caused by solid particles or liquid droplets impacting the tooth surface, adhesive wear resulting from direct contact and friction between mating teeth, and abrasive wear due to hard particles or contaminants in the lubricant. These wear mechanisms can occur independently or in combination, leading to surface degradation, increased material loss, and altered gear geometry, all of which can impair gear function. Multiple factors commonly associated with wear in gear systems include inadequate lubrication, abrasive contaminants, misalignment, high operating temperatures, excessive loads, and environmental conditions. Figure 1.2 illustrates gear failure through wear.

Figure 1.2. Gear failure through wear

1.2.2. Tooth breakage

When gear teeth experience high pressures, sudden shocks, or impact forces, they can fracture, chip, or deform, resulting in tooth breakage, a severe and critical form of gear failure. This type of failure can occur suddenly and catastrophically, causing significant damage, system downtime, and potential safety risks to gear systems. Tooth breakage often results from overloading, abrupt shocks, material defects, inadequate design considerations, manufacturing errors, and operational irregularities. These factors can create excessive stress concentrations, surpassing the material's strength limits and compromising the tooth's ability to withstand loads. Prompt identification, intervention, and remedial action are essential to restore gear system functionality and prevent further damage or catastrophic failure. Consequences of tooth breakage include impaired gear function, loss of motion transmission, system inefficiencies, increased noise and vibration levels, and possible damage to surrounding components. There are numerous intricate and interconnected reasons behind tooth breakage in gear systems, including inadequate design, inferior material quality, manufacturing flaws, operational misuse, and environmental conditions that induce excessive stress, fatigue, and the initiation of tooth failure. Figure 1.3 illustrates gear tooth breakage.

Figure 1.3. Gear tooth breakage

1.2.3. Fatigue

Fatigue, a significant mode of gear failure, occurs due to repeated cyclic loading on gear teeth during operation, leading to progressive degradation, initiation, and propagation of fractures. The cumulative effect of cyclic loading characterizes this failure mode, subjecting the gear material to alternating stresses and strains beyond its endurance limit. This cyclic loading induces the development and propagation of small cracks or flaws, particularly at stress concentration cites like tooth root fillets and gear tooth surfaces. Fatigue failure in gear systems often manifests as crack formation, resulting in wear, surface pitting, tooth fracture, and spalling, ultimately diminishing gear life. Figure 1.4 illustrates the fatigue failure.

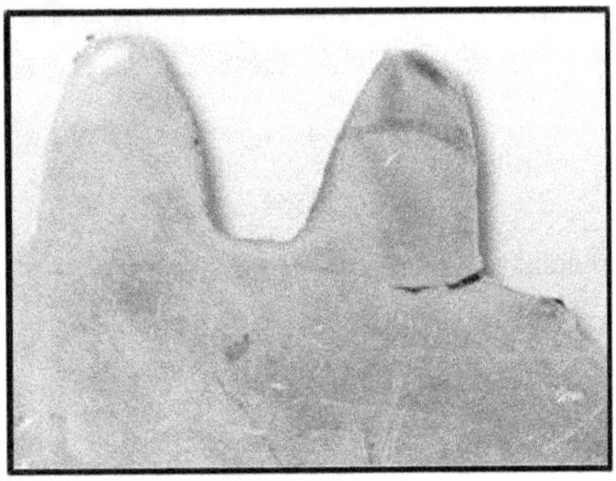

Figure 1.4. Fatigue failure

1.2.4. Pitching

Pitting, a common and detrimental form of gear failure, occurs when localized pits or craters develop on the surface of gear teeth due to surface fatigue and material removal during operation. This failure mode is typically attributed to repetitive cyclic loads, inadequate lubrication, lubricant impurities, and material characteristics that lead to micro-spalling, fracture initiation, and material loss on the tooth surface. Pitting often occurs at high-stress concentration areas such as the tooth contact zone when applied loads exceed the material's endurance limit, resulting in surface distress, microcracks, and pit formation due to material fatigue and degradation. Failure to promptly address pitting in gear systems through preventive maintenance and corrective actions may compromise the system's performance. Consequences of pitting in gear systems include increased friction, elevated operating temperatures, noise, vibration, reduced gear efficiency, and potential tooth breakage or surface wear. Figure 1.5 illustrates the gear failure through pitting.

Figure 1.5. Pitting failure

1.2.5. Scoring

Scoring, a detrimental form of gear failure, arises from frictional forces and mechanical interactions during operation, resulting in localized scratches, grooves, or marks on the gear tooth surface. Common causes of this failure mode include high operating temperatures, lubricant impurities, misalignment between mating gears, and material characteristics leading to surface distress or abrasive/adhesive wear at the tooth contact zone. Areas with high friction, such as tooth flanks or root fillets, are particularly susceptible to scoring due to surface damage, material loss, and roughening caused by direct contact, friction, and metal-to-metal interactions. These factors reduce gear system efficiency. Addressing scoring in gear systems is crucial to restoring functionality, preventing further damage, and optimizing performance across various industrial applications and operating conditions. The adverse effects of scoring include increased friction, elevated temperatures, noise, vibration, reduced efficiency, accelerated wear, and potential tooth breakage, necessitating prompt identification, intervention, and remediation. Figure 1.6 illustrates the gear failure through scoring.

Figure 1.6. Scoring failure.

1.2.6. Plastic flow

Under extreme loads, stress concentrations, or high temperatures during operation, gear teeth can undergo permanent deformation or yield, a significant and troublesome mode of gear failure known as plastic flow. This failure mechanism typically occurs due to overloading, sudden shock loads, material softening induced by elevated temperatures, inadequate material strength, or insufficient design considerations, leading to excessive stresses, strain rates, and localized plastic deformation in the gear material. Plastic flow often manifests when applied stresses surpass the material's yield strength, particularly at high-stress concentration sites like tooth root fillets, gear tooth surfaces, or contact zones, resulting in plastic deformation, alterations to gear shape, and diminished gear functionality. Plastic flow in gear systems can lead to changes in gear profiles, misalignment, poor meshing, increased friction, noise, vibration, reduced efficiency, and even tooth breakage or failure. Consequently, it is vital to promptly identify, intervene, and rectify the issue to preserve gear system functionality, prevent further damage, and enhance gear performance across diverse industrial applications and operating conditions. Figure 1.7 illustrates the gear failure through excessive heating (i.e., plastic flow).

Figure 1.7. Tooth deformation due to excessive heat

1.2.7. Thermal cycle fatigue

Thermal cycle fatigue is a critical mode of gear failure, resulting from repeated thermal loads and mechanical strains induced by cyclic temperature fluctuations during gear operation. This failure mechanism is typically caused by rapid temperature changes, thermal gradients, and the expansion and contraction of gear components, which lead to varying mechanical strains, thermal stresses, and material fatigue in the gear material. When cyclic temperature variations exceed the material's thermal fatigue resistance, thermal cycle fatigue often occurs at high-temperature gradients, stress concentration areas, or heat-affected zones such as tooth root fillets, gear tooth surfaces, or contact zones. Under cyclic thermal and mechanical loading conditions, this can initiate cracks, propagate them, and eventually lead to gear failure. It is essential to promptly identify, intervene, and remediate its consequences to address thermal cycle fatigue and enhance gear system performance across diverse industrial applications and operating conditions. Figure 1.8 illustrates the gear failure thermal cycle fatigue.

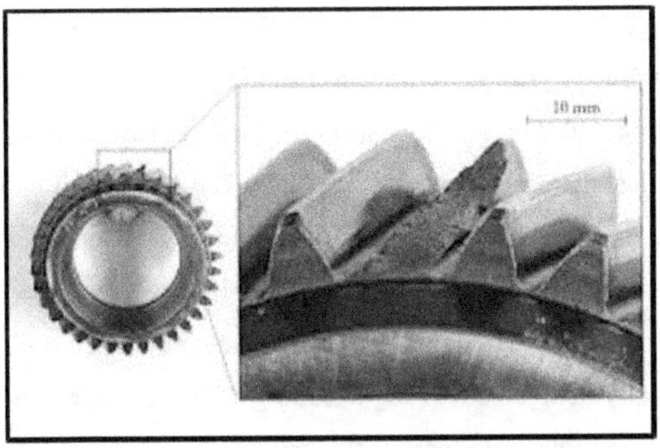

Figure 1.8. Thermal cycle fatigue of gear

1.3. Stresses in gears

Gears are vital mechanical components that transmit power and motion between rotating shafts and play a crucial role in equipment functionality. Understanding gear action is crucial before delving into the stresses that develop in plastic gears. When the driver and driven gears come into contact, they tend to bend due to the load. Therefore, gears must possess both high flexural strength and stiffness. Friction forces induce bending stresses and create other stresses on the tooth surface. This can be observed as either line or point contact between the meshing gear surfaces. Gear tooth meshing involves an initial load placed on one another, resulting in internal compressive stresses on the gear surface during point contact. Additionally, frictional forces come into play when one gear surface slides over another. This force generates tensile stress in the gear when subjected to line contact. In summary, gear meshing involves continuous point and line contact between the two mating gears. In Figure 1.9a, 'R' represents the rolling position, and 'S' represents the sliding position. When these two force directions are opposite, problems arise. In the depicted gear meshing scenario, the gear material experiences compression during rolling (observed at Point 1) and tension during sliding (identified just away from the pitch point). Consequently, the gear surface may suffer from cracking, surface fatigue, and the formation of build-up edges, leading to material erosion. When gears mesh, they should roll in the same direction toward the pitch point (Point 2). The mating gears exhibit a compression state during rolling and a tension state during sliding (Point 3).

When two gears mesh, the rolling motion remains in the same direction, but the sliding motion changes direction. It becomes evident that the driven gear experiences tension during sliding and compression during rolling. For the driving gear, the tip of the tooth encounters less damage or stress at Point 5 (compression) and Point 6 (tension). At the pitch point, due to sliding forces, pure rolling occurs—a null point for sliding is created because of the directional change. The susceptibility to surface failure is low in this region since it is not subjected to compound stresses near the pitch point. However, serious failure can occur when subjected to high loading. This reduction in load distribution happens between the previous tooth pair or the next tooth pair during initial gear contact (as shown in Figure 1.9a) or at the end of contact (Figure 1.9b), slightly above the pitch line at the pitch point. In such cases, one tooth pair bears most or all of the load, leading to severe heat build-up and surface deterioration. Ultimately, this can result in fatigue failure [23–25].

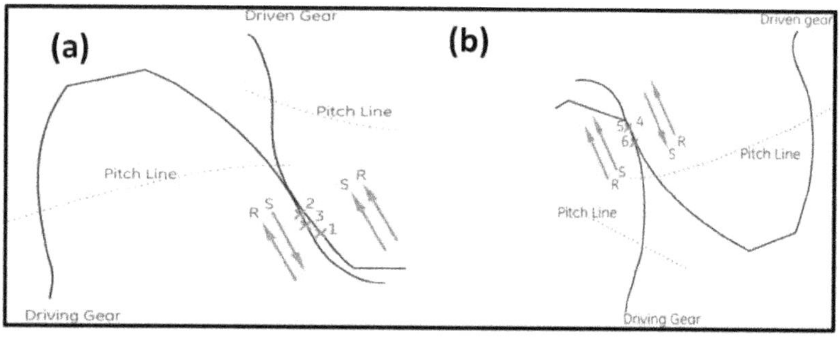

Figure 1.9. (a) Beginning of gear tooth contact and (b) Ending of gear tooth contact

Designing reliable gear systems necessitates a deep comprehension of the stresses they encounter during operation. The intricate interplay among mating teeth, applied loads, operational parameters, material characteristics, and geometric aspects gives rise to diverse stresses on gears, encompassing torsional, bending, and contact stresses.

1.3.1. Contact stresses

The primary stresses generated during operation at the interfaces where gear teeth mesh are called contact stresses. These stresses arise from the normal and tangential forces transferred between the contacting tooth surfaces, leading to shear stresses, surface interactions, and localized pressure distributions. The magnitude and distribution of these contact stresses are influenced by various factors, including applied loads, tooth geometry, profile adjustments, material properties, lubrication, and operating conditions. It is imperative to employ appropriate design, material selection, lubrication, and maintenance practices to enhance gear contact performance, mitigate stress concentrations, and enhance gear efficiency. Excessive contact stresses can result in surface fatigue, pitting, scoring, wear, and potential tooth breakage or failure [2, 3, 28, 8, 8, 10, 16, 19, 22, 26, 27].

1.3.2. Bending stresses

Bending stresses on gear teeth arise from forces and bending moments perpendicular to the tooth profile during operation. These stresses induce bending deformation, tooth deflection, and stress concentrations, particularly at the tooth root fillets and gear tooth surfaces. They include applied loads, misalignments, clearances, tooth geometry, profile adjustments, and operational conditions. Bending stresses exhibit varying magnitudes and distributions across the tooth profile, peaking at the fillets of the tooth root and diminishing towards the tooth tips. To mitigate stress concentrations, enhance tooth strength, durability, and resistance to bending fatigue, and ensure the reliability and longevity of gear systems across diverse industrial applications and operating conditions, it is essential to consider safety factors, optimize tooth geometry, and select suitable materials. Excessive bending stresses can also lead to tooth bending fatigue, initiation and propagation of cracks, and eventual tooth breakage or failure [29, 29].

1.3.3. Torsional Stresses

During gear operation, forces and torsional moments along the gear tooth profile give rise to torsional stresses within the gears. These stresses induce torsional deformation, shear stresses, and stress distributions across the tooth profile, influenced by factors such as applied torques, rotational speeds, gear ratios, misalignments, clearances, and operating conditions. Various parameters, including torque distribution, tooth geometry, profile adjustments, material properties, and lubrication, impact the magnitude and distribution of torsional stresses. To minimize stress concentrations, bolster tooth strength, durability, and resistance to torsional fatigue, and ensure the reliability of gear systems across diverse industrial applications and operating environments, attention must be given to safety factors, optimization of tooth geometry, and selection of materials. Excessive torsional stresses can lead to shear fatigue, material degradation, and potential tooth failure [2, 5, 8].

1.4. Selection criteria for gear materials

Selecting suitable materials for gears is crucial for maximizing their life and performance. Material selection should consider various factors, including environmental conditions and mechanical properties. Key considerations in choosing gear materials include strength, coefficient of friction, operating temperature, chemical exposure, precision requirements, and compatibility with mating gear materials [15, 28, 30–33].

1.4.1. Accuracy

In industries demanding precision and accuracy, such as robotics, automotive, and aerospace, achieving the required gear precision is paramount. Material selection plays a crucial role in attaining this accuracy, with the material's mechanical properties, manufacturing consistency, and dimensional stability being key factors. Materials characterized by a uniform microstructure, low internal stresses, and exceptional dimensional stability, such as premium alloy steels, tool steels, and specialized high-performance metals, are highly recommended for high-precision gear applications. These materials offer superior gear performance and precision, particularly in applications necessitating tight tolerances, precise positioning, and meticulous motion control. Additionally, they ensure consistent mechanical properties, enhanced geometric integrity, improved gear accuracy, minimized dimensional variations, and reduced gear errors.

1.4.2. Mating gear material

The reliability of the system, wear characteristics, frictional behavior, and overall gear performance are significantly influenced by the compatibility and tribological properties of the mating gear materials. To achieve optimal load distribution, proper mating, efficient power transfer, and minimal wear, friction, and surface damage during operation, it is essential to pair materials appropriately. Superior gear performance, reliability, and compatibility across various applications and operating conditions are ensured by materials with complementary mechanical properties, surface hardness, wear resistance, and coefficient of friction. Examples of such materials include carburized steels paired with nitride steels, surface-treated steels paired with bronze alloys, or high-performance alloys combined with specialized coatings.

1.4.3. Chemical environment

In corrosive, aggressive, and hostile operating environments, the selection of materials, compatibility, and performance is influenced by factors such as the chemical surroundings, exposure to aggressive substances, susceptibility to chemical attack, corrosion, oxidation, and environmental degradation. In such challenging chemical conditions, it is essential to use materials that exhibit superior chemical resistance, corrosion resistance, and environmental stability to maintain material integrity, protect surfaces, and ensure sustained functionality. Stainless steels, corrosion-resistant alloys, surface-treated metals, protective coatings, and specialized high-performance materials are examples of materials with excellent chemical resistance. These materials offer enhanced material stability, reduced chemical degradation, fewer failures due to corrosion, and longer gear life, thereby contributing to optimal gear performance and longevity in demanding chemical environments.

1.4.4. Service temperature

In scenarios involving high temperatures, high speeds, and thermal cycling, the performance of gears, thermal management, dimensional stability, and material integrity are significantly influenced by factors such as the material's service temperature, thermal stability, heat resistance, thermal expansion, and thermal fatigue resistance. Materials like heat-treated steels, alloy steels, high-temperature alloys, and specialized high-performance materials exhibit notable attributes such as high thermal stability, resistance to heat, low coefficients of thermal expansion, and excellent resistance to thermal fatigue. Additionally, these materials minimize dimensional changes, enhance material stability, facilitate efficient thermal management, reduce thermal stresses, and contribute to prolonged gear life. Consequently, they ensure optimal gear performance, reliability, and longevity in applications characterized by high temperatures, heavy loads, and intensive-duty conditions.

1.4.5. Strength

Material strength, along with dynamic stresses and changing operating conditions, collectively influence various gear failure modes, fatigue resistance, shock resistance, strength, hardness, toughness, and fatigue strength. In demanding applications characterized by high loads, high speeds, and heavy-duty requirements, materials possessing high strength, hardness, toughness, and fatigue resistance, such as alloy steels, tool steels, high-performance metals, or specialized high-strength materials, offer enhanced shock resistance, prolonged fatigue life, reduced risk of failure, and optimized gear performance and durability.

1.4.6. Coefficient of friction

Gear mating, tooth contact, power transmission, energy efficiency, wear characteristics, and system reliability are significantly influenced by factors such as the material's coefficient of friction, surface finish, lubrication requirements, and frictional behavior. Materials like surface-treated steels, nitrided steels, self-lubricating alloys, or specialized friction-reducing materials are examples of materials characterized by low coefficients of friction, high surface hardness, excellent wear resistance, and superior tribological properties. These materials also contribute to reduced frictional losses, improved gear mating, optimized power transmission, prolonged gear life, and enhanced system efficiency, thereby ensuring optimal gear performance, as well as energy efficiency across various applications and operating conditions.

1.5. Gear materials

Gears are crafted from a variety of materials selected based on their specific attributes, compatibility with application requirements, and environmental considerations. The choice of material plays a crucial role in determining equipment performance, lifespan, and operational efficiency. Below outlines the primary materials utilized in gear manufacturing, along with their respective characteristics [7, 25, 31, 33]:

1.5.1. Metallic gears

A variety of metals and metal alloys, including steel, brass, bronze, aluminum, and titanium, are utilized to manufacture metallic gears. Metallic gears, serving as crucial components in numerous mechanical systems and machinery across various sectors, offer reliable power transfer, motion control, and mechanical advantage due to their inherent strength, durability, and versatility.

Steel Gears: Among the most versatile components in mechanical engineering and industrial applications, steel gears hold the inherent qualities of steel to deliver a unique blend of performance, robustness, and longevity. Their exceptional strength and load-bearing capacity are particularly notable, enabling them to withstand heavy loads, shocks, and impacts without deformation or failure. Steel gears excel in harsh environments, making them indispensable for heavy-duty applications in industrial machinery, construction equipment, and automobile transmissions, among others. In addition to their strength, steel gears demonstrate outstanding resilience and wear resistance, ensuring extended service life and reduced maintenance requirements. Various heat treatment methods, such as carburizing, nitriding, and induction hardening, enhance the wear resistance, fatigue strength, and overall performance of steel gears by modifying their microstructure and increasing surface hardness.

Carbon steel gears: Widely used and cost-effective in mechanical engineering and industrial applications, carbon steel gears offer affordability, reliability, and durability due to their high strength and moderate hardness. Notably, the intrinsic strength of carbon steel gears, derived from the carbon content in the steel alloy, provides them with high toughness and tensile strength. As a result, carbon steel gears can withstand heavy loads, shocks, and impacts without deformation or breakage, making them suitable for general-purpose applications requiring support for moderate to heavy loads. While exhibiting moderate hardness under typical operating conditions, carbon steel gears offer sufficient resistance against wear, abrasion, and surface fatigue. Additionally, they can be further hardened through heat treatment methods like carburizing or induction hardening to enhance wear resistance and extend service life in increasingly demanding environments.

Stainless steel gears: Renowned for their exceptional durability, corrosion resistance, and precision performance, stainless steel gears represent a distinct class of gears ideal for demanding and harsh operating conditions. Leveraging the unique properties of stainless-steel alloys, these gears offer a dependable and long-lasting solution for applications requiring resistance to rust, corrosion, and environmental degradation. The chromium concentration in stainless steel alloys creates an oxide layer that protects the material's surface, resulting in unmatched resistance to oxidation and corrosion. Consequently, stainless steel gears can withstand moisture, chemicals, saltwater, and other corrosive agents, ensuring prolonged service life, reduced maintenance, and enhanced reliability compared to carbon steel or alloy steel gears. Additionally, stainless steel gears exhibit high tensile strength and toughness, enabling them to withstand heavy loads, shocks, and impacts without deformation or fatigue, making them suitable for heavy-duty applications in industries such as marine, offshore, chemical processing, food and beverage, and pharmaceuticals.

Cast iron gears: Cast iron gears, crafted from cast iron alloys known for their strength, stability, and affordability, offer a dependable and cost-effective solution for various mechanical and industrial applications requiring durability, strength, and stability. Characterized by exceptional resistance to deterioration, abrasion, and distortion, cast iron gears feature a microstructure primarily composed of silicon, ferrite, iron, and graphite flakes embedded in a matrix. This unique microstructure contributes to excellent wear resistance, hardness, and compressive strength, enabling cast iron gears to resist deformation and failure under heavy loads, shocks, and impacts. Consequently, they are well-suited for heavy-duty applications in industrial gear, construction equipment, agricultural machinery, and other machinery. Additionally, cast iron gears exhibit outstanding thermal stability, thermal deformation resistance, and expansion resistance under temperature fluctuations, ensuring consistent performance and dimensional stability across a wide temperature range. Their intrinsic thermal conductivity and heat-dissipating capabilities help mitigate the negative effects of thermal stresses, guaranteeing reliable operation in high-temperature environments, fluctuating conditions, or thermal cycling scenarios commonly encountered in heavy machinery, power transmission systems, and engines.

1.5.2. Non-metallic gears

Polymer gears: These gears, crafted from thermoplastic and thermosetting polymer materials such as nylon, polyethylene, polyacetal (Delrin), and polyetherketone (PEEK), offers a blend of affordability, adaptability, and lightweight design. They find applications across various industries, including consumer electronics, automotive parts, medical equipment, and industrial machinery. Lightweight construction is a key benefit of plastic gears, aiding in applications where material efficiency and weight reduction are crucial. This leads to reduced inertia, lower energy consumption, and increased efficiency. Plastic gears are also valued for their low cost and high productivity, as they can be mass-produced with complex geometries and tight tolerances using plastic molding and extrusion techniques.

Ceramic gears: These gears, made from ceramic materials like silicon nitride (Si_3N_4), alumina (Al_2O_3), zirconia (ZrO_2), and silicon carbide (SiC), are well-suited for high-performance applications in industries such as precision manufacturing, automotive, aerospace, and medicine. With superior mechanical properties, wear resistance, temperature stability, and corrosion resistance, ceramic gears excel in high-load, high-speed, and abrasive conditions. They maintain consistent performance, minimal tooth wear, and reduced frictional losses due to their unique crystalline structure and chemical composition. Additionally, ceramic gears offer thermal stability, resistance to thermal expansion, and resistance to temperature variations, making them ideal for applications exposed to extreme temperatures and thermal cycling.

1.6. Composites as a gear material

Composite gears, a specialized category of gears crafted from composite materials, typically consist of a matrix composed of epoxy, polyester, or polyimide resin combined with carbon, glass, or aramid fibers. These innovative materials find extensive application in high-performance and specialized sectors across industries like aerospace, automotive, marine, and sporting goods. Their unique blend of lightweight design, high tensile strength, stiffness, and tailored performance characteristics makes them ideal for various applications.

One of the primary advantages of composite gears is their lightweight construction, which addresses the need for weight reduction, material efficiency, and performance optimization in machinery and systems. By reducing inertia, consuming less energy, and enhancing efficiency, they contribute significantly to improved operational outcomes. Additionally, their lower density compared to traditional metal alloys makes them well-suited for applications where weight reduction, fuel efficiency, and performance enhancement are paramount. Examples of such applications include aerospace propulsion systems, automotive drivetrains, and sporting equipment. Furthermore, composite materials enable the creation of lightweight and compact gear assemblies without compromising mechanical integrity, strength, or durability.

Composite gears are also valued for their rigidity, high tensile strength, and specific performance attributes. Tailoring the volume fraction, type, and orientation of reinforcing fibers within the composite matrix allows for the optimization of desired mechanical properties such as strength, stiffness, fatigue, and wear resistance, enabling them to outperform traditional metal gears in high-stress, high-speed, and abrasive environments. Applications in racing, aerospace components, and industrial machinery requiring resilience in harsh operating conditions benefit greatly from composite gears' adaptability and customizability.

Additionally, composite gears exhibit strong performance qualities, high strength, and lightweight characteristics. They offer excellent resistance to corrosion, thermal stability, and environmental factors owing to their inherent chemical inertness and non-reactive nature. This ensures long-term performance in corrosive, acidic, or caustic environments such as those found in chemical processing, maritime applications, and industrial machinery exposed to saltwater or aggressive chemicals. Moreover, their exceptional thermal stability and resistance to thermal expansion make them well-suited for thermal cycling, rapid temperature changes, and harsh environments, including aerospace parts, automobile engines, and industrial machinery [7, 31–33].

1.6.1. Aluminum as a gear material

Apart from aluminum, metallic gears often encounter challenges such as susceptibility to corrosion, increased weight leading to reduced efficiency, and the potential for amplified noise and vibration. Additionally, these gears may require frequent maintenance due to wear and tear, thereby escalating manufacturing and replacement costs. On the other hand, non-metallic gears, such as those composed of plastic and composite materials, may lack the necessary mechanical robustness and longevity required for high-load and high-speed applications, resulting in shortened lifespans and diminished dependability. Furthermore, they may be prone to deformation or failure under harsh conditions due to temperature changes. Consequently, they may necessitate specific treatments and careful material selection to enhance performance and longevity, given their potential vulnerability to chemical exposure and abrasive wear.

However, aluminum gears are gaining popularity in mechanical engineering and industrial applications owing to their exceptional mechanical properties, lightweight nature, and versatility. Being a non-ferrous metal, aluminum offers distinct advantages over traditional steel gears, particularly in scenarios where corrosion resistance, weight reduction, and material efficiency are crucial considerations. Aluminum proves to be an ideal material for situations demanding weight reduction without compromising strength and performance, given its lightweight composition with a density approximately one-third that of steel. Furthermore, aluminum exhibits exceptional mechanical properties, including outstanding corrosion resistance, high tensile strength, and excellent machinability, especially when appropriately coated or treated.

The utilization of aluminum alloys, such as 6061-T6 and 7075-T6, in gear production is widespread due to their well-balanced strength, hardness, fatigue resistance, and toughness. As a result, aluminum gears can withstand heavy loads, shocks, and impacts over extended periods without compromising dimensional stability or performance consistency. Moreover, the ease of machining aluminum alloys enables the precise manufacturing of gears with intricate profiles and tight tolerances, allowing for the creation of specialized designs tailored to specific application requirements and performance standards. The aluminum gears are shown in Figure 1.10 [34, 35].

Figure 1.10. Aluminum gears

1.6.2. Advantages of aluminum gears

- The primary advantage of aluminum gears lies in their lightweight design, which is beneficial for machinery and systems prioritizing weight reduction and material efficiency. This feature reduces inertia, lowers energy consumption, and enhances overall efficiency.
- Aluminum demonstrates excellent thermal conductivity and heat dissipation capabilities, facilitating efficient heat transfer and dissipation. This property aids in mitigating the negative impacts of thermal stresses, temperature fluctuations, or high-temperature operating conditions.
- Aluminum is abundant, widely accessible, and cost-effective compared to other high-performance materials like titanium or composites. As a result, it represents an economical and efficient choice for gear manufacturing, particularly for mass-produced consumer goods, automotive components, household appliances, and industrial machinery.
- Due to the formation of a protective oxide layer on its surface, aluminum exhibits inherent corrosion resistance, minimizing the effects of environmental exposure, moisture, and oxidation. These characteristics, combined with its mechanical performance, thermal stability, and lightweight construction, make aluminum gears suitable for a wide range of applications, as shown in Figure 1.11 [7, 31, 36].

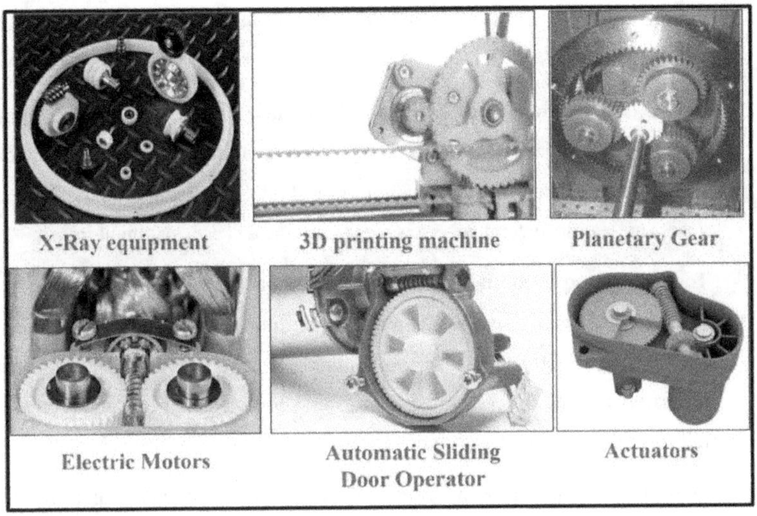

Figure 1.11. Applications of aluminum gears in various sectors

1.6.3. Pitfalls of aluminum gears

While aluminum gears offer advantages such as lightweight construction, corrosion resistance, and thermal stability, they also come with drawbacks that may impact their reliability and performance, especially when compared to steel gears. The primary limitations of aluminum gears, including lower strength, poor wear resistance, and a low elastic modulus, contribute to significant tooth deflection and several associated disadvantages outlined below [7, 25, 31, 32]:

Reduced strength: Aluminum alloys generally exhibit lower tensile and yield strengths compared to steel or alloy steel. Consequently, they are more susceptible to deformation, yielding, and failure under high-load, high-stress, or impact conditions. This decreased strength and load-bearing capacity may lead to premature wear, fatigue, or failure in applications where toughness, resilience, and strength are critical.

Limited wear resistance: In environments characterized by high speeds, heavy loads, or abrasive conditions, aluminum gears are more prone to abrasive wear, adhesion, galling, and surface fatigue due to their lower wear resistance and surface hardness compared to steel or alloy steel gears. This reduced wear resistance and surface hardness can accelerate wear, tooth thinning, and surface degradation, necessitating more frequent maintenance and shortening gear life.

Low elastic modulus and high tooth deflection: The lower elastic modulus of aluminum alloys makes them prone to bending, flexing, and deformation under load, torque, or thermal stresses compared to steel or alloy steel. Significant tooth deflection in aluminum gears can lead to several severe issues and failure scenarios, including:

Tooth thinning from high wear: Excessive tooth deflection in aluminum gears, combined with reduced wear resistance, can accelerate tooth abrasion, thinning, and wear. This can alter gear geometry, decrease tooth thickness, and compromise load-bearing capacity, ultimately leading to accelerated wear rates, shortened gear life, and increased risk of tooth failure and gear tooth fracture over time.

Tooth deformation due to heat: The low thermal conductivity and high tooth deflection of aluminum alloys can result in thermal strains, localized heating, and excessive heat generation within gear teeth, especially at high speeds, heavy loads, or prolonged operation. This heat buildup can cause tooth deformation, altered gear geometry, and diminished mechanical properties due to thermal expansion, distortion, and localized softening. Overheating-induced tooth deformation can exacerbate wear rates, reduce gear life, and elevate the risk of tooth fracture and failure in thermal cycling or high-temperature environments.

Tooth fracture from overloading: High tooth deflection, combined with lower strength and fatigue resistance relative to steel or alloy steel, can lead to tooth fracture, splitting, and failure in aluminum gears subjected to overloading, shock loading, or impact. The reduced fatigue strength, fracture toughness, and load-bearing capacity of aluminum gears may result in catastrophic gear failure, tooth fracture, and potential damage to associated systems or machinery, causing operational disruptions, safety hazards, and costly repairs or replacements.

1.7. Aluminium metal matrix nanocomposites

Nanocomposites have revolutionized gear technology by combining the inherent advantages of aluminum with the benefits of nanoscale reinforcements, resulting in

Aluminum Metal Matrix Nanocomposites (AMMNCs). Through the integration of nanoparticles, AMMNCs achieve superior mechanical properties such as enhanced stiffness, wear resistance, hardness, and strength, marking a significant advancement in gear performance and durability. AMMNCs consist of an aluminum matrix embedded with nanoscale reinforcements like nanofibers or nanoparticles distributed evenly throughout. This strategic dispersion of reinforcements creates a synergistic effect that elevates the composite's mechanical and physical characteristics. Engineers can tailor AMMNCs to specific applications by precisely adjusting the size, concentration, and dispersion of nanoparticles, optimizing performance across diverse operational environments.

Incorporating nanoparticles into the aluminum matrix substantially enhances the tensile and compressive strength of AMMNCs. The high aspect ratio and surface area of nanoparticles facilitate effective load distribution, hindering dislocation movement and plastic deformation within the aluminum matrix. This heightened strength equips AMMNCs with improved load-bearing capacity, enhanced fatigue resistance, and reduced susceptibility to premature failure in challenging operating conditions. Furthermore, nanoscale reinforcements significantly bolster the toughness of AMMNCs, enhancing their resistance to surface fatigue, wear, and abrasion. The efficient dispersion of load by the reinforcements mitigates localized wear and prevents surface deformation, thereby extending gear life, reducing maintenance intervals, and enhancing operational reliability, particularly in scenarios involving high loads, elevated speeds, and abrasive environments. Various nanoparticles can be added as reinforcements to aluminum to prepare AMMNCs. Figure 1.12 shows the various reinforcements and their properties [37–39].

Reinforcements / 50 nm	Density (g/cc)	Fracture toughness (MPa.m½)	Structure	Hardness (Kg / mm²)	Melting point (°C)	Coefficient of thermal expansion (°C⁻¹)	Thermal conductivity (W/m.°C)	Cost in US Dollars (100grams) (US Research nanomaterials)
Al_2O_3	3.9	4.5	Trigonal	2100	2072	8.4×10^{-6}	35	98
AlN	3.26	2.6	Hexagonal	1100	2200	4.6×10^{-6}	320	188
B_4C	2.52	3	Hexagonal	2800	2763	5×10^{-6}	35	3700
BN	2.29	2.6	Hexagonal	205	2973	38×10^{-6}	1700	298
Carbon nanotubes	1.6	2.7	Cylindrical	2600	2800	2×10^{-5}	3500	685
Graphene dry powder	2	2.8	Hexagonal	-	4627	-	550	17160
SiCp	3.218	4	Cubic	2500	2730	5.12×10^{-6}	120	88
SiO_2	2.4	0.62	Tetragonal	800	2230	0.65×10^{-6}	1.3	55
Graphite	2.23	2.2	Hexagonal	1700	4027	4.06×10^{-6}	398	149
Si_3N_4	3.2	6.1	Trigonal	1580	1900	3.3×10^{-6}	30	198
TiC	4.93	-	Cubic	2470	3160	7×10^{-6}	5.64	145
TiN	5.2	-	Cubic	2100	2930	9.35×10^{-6}	19.2	168
TaC	14.4	-	Cubic	1700	3880	-	21	1900
TiB_2	4.52	6.2	Hexagonal	1800	3230	6.4×10^{-6}	25	9800
WC	15.6	-	Hexagonal	2100	2870	1×10^{-6}	110	108
WO_3	7.16	-	Monoclinic	720	1473	-	-	126

Figure 1.12: Reinforcements and their properties

1.8. Importance and motivation of the study

The present study holds significant importance in the field of mechanical engineering and materials science. In a world where mechanical systems are continuously evolving to meet the demands of various industries, understanding the behavior of gear systems under different conditions is crucial for enhancing efficiency, reliability, and life. This investigation addresses various important aspects that contribute to advancing the understanding and optimization of gear performance, particularly when utilizing novel composite materials like aluminum-based hybrids. One of the primary reasons for the importance of this study lies in the widespread use of spur gears across various applications, ranging from automotive transmissions to industrial machinery and power generation systems. Spur gears are among the most commonly employed gears due to their simple construction, cost-effectiveness, and reliability. However, their performance can be significantly affected by factors such as misalignment, which can lead to increased wear, noise, and reduced efficiency. By investigating the contact stresses induced in spur gears under misalignment conditions, this investigation aims to provide valuable insights into the behavior of these gears in real-world scenarios.

Moreover, using optimally developed aluminum-based hybrid composites introduces an innovative aspect to the study. Composite materials offer unique advantages such as high strength-to-weight ratio, corrosion resistance, and tailored mechanical properties. By incorporating aluminum-based hybrids into gear manufacturing, engineers can potentially enhance the performance and life of gear components while reducing weight and improving energy efficiency. However, the interaction between these composite materials and the operational conditions of gears, particularly under misalignment, remains a relatively unexplored area. Therefore, this investigation fills a crucial gap in the current understanding of gear materials and their behavior under challenging operating conditions.

1.9. Problem statement

The development of various hybrid composites, such as combinations of silicon carbide (SiC) with aluminum oxide (Al_2O3), SiC with aluminum, titanium metal powder with graphite, and nano-alumina with aluminum 6061 (Al6061) and molybdenum disulfide (MoS_2), as well as hybrids like aluminum 7150 with hexagonal boron nitride (hBN), underscores ongoing efforts to enhance aluminum-based materials by integrating diverse reinforcements. However, challenges remain in using aluminum matrix composites for gear production as an alternative to traditional metal gears. Key issues include inadequate bonding strength between the matrix and reinforcements, which can diminish mechanical performance and high wear rates that affect gear durability and reliability. Current research focuses on creating hybrid nanocomposite gears to improve bond strength and wear resistance. The goal is to refine composite formulations and processing methods to produce gears with superior performance and durability, making them viable alternatives to conventional metal gears. This approach aims to address existing limitations and leverage the advantages of hybrid nanocomposites for advanced engineering applications.

Engineers aim to achieve cheap, low-weight, highly durable solutions while maintaining maximum design perfection. Gears, fundamental in everyday life, must be high-quality to manage contact and restrict contact stress, especially in high-speed applications. For centuries, gears of various shapes and dimensions have been used in all sectors, and for the last two decades, they have been the most studied element, with ongoing research. It is necessary to develop and produce gearboxes with low weight, precision, and highly durable gears. These gears should be made from high-strength, low-weight materials capable of carrying maximum loads, transferring these loads in high-velocity environments with low vibrations and noise. They must resist fatigue wear stress, maintain structural capability at high temperatures, and have surface resistance to pitting wear rate. Misalignment in gears can increase contact stress, wear, and noise and lead to gear failure, yet it has not received much attention from the research community. Reducing misalignment in gear design and operation is essential. A comprehensive literature review reveals that most authors use simulation methods to study gear performance under misalignment, with few employing experimental approaches. Thus, this work follows a dual approach, using simulation and experimental investigation to evaluate contact stress in aluminum composite gears.

1.10.Organization of thesis

The thesis was divided into five chapters: Introduction, Literature review, Objective-1, Objective-2, and Conclusions.

Chapter 1: This chapter explains the importance of gear materials, focusing on aluminum metal matrix nanocomposites. It addresses gear failures, contact stresses, and criteria for selection of materials, emphasizing the suitability of aluminum composites for gear applications. This chapter also highlights the challenges and motivations behind using aluminum nanocomposites, offering a strong foundation for further research in optimizing gear performance through innovative material choices.

Chapter 2: This chapter deals into the fabrication of composites, investigates studies on contact stresses, and discusses relevant research related to the current study. It contains a comprehensive summary of existing literature while identifying gaps in knowledge. The chapter effectively sets the stage for the proposed research by outlining clear objectives to address the identified gaps in composite and stress analysis.

Chapter 3: This chapter thoroughly explains mechanical and tribological characteristics in Al 6061 hybrid nanocomposites improved with SiC, B_4C, and GO. Through a novel optimized mix selection approach, it effectively demonstrates enhancements in material performance.

Chapter 4: This chapter effectively combines simulation and experimental testing to analyze contact stress in aluminum composite spur gears. It provides a complete examination of gear performance under misalignment, offering valuable insights for enhancing gear durability and precision in practical applications.

Chapter 5: This chapter contains the summary of conclusions and future scope of the present work with possible remarks and future scope.

References

Publications

CHAPTER-2

2. LITERATURE REVIEW

2.1. Introduction

Composite materials are created by combining two or more constituent materials, each with distinct physicochemical properties [40]. Adding reinforcement materials enhances the stiffness and strength of the base material; typically, the reinforcement is stronger and more rigid than the matrix [41]. The continuous phase matrix can incorporate reinforcing materials, such as fibers, particles, or whiskers [42]. The effectiveness of these reinforcements dependent on their morphology and distribution within the composite [43]. Reinforcement materials can be either discontinuous or continuous. Metal Matrix Composites (MMCs) are known for their low density, high strength, and higher modulus of elasticity. Due to these enhanced properties, MMCs are used in a broad range of applications. When a composite includes multiple forms of reinforcement within the matrix, it is called a Hybrid Metal Matrix Composite (HMMC) [44]. Specifically, composites resulting from reinforcing aluminum alloy matrices with materials like boron carbide (B_4C) or silicon carbide (SiC) are known as Particulate Metal Matrix Composites (PMMCs) [45–47]. These composites exhibit exceptional wear resistance and significant strength due to their strong particle reinforcements.

Hybrid nanocomposites are increasingly favored in biomedicine for enhancing the tribological properties of medical devices [48]. Their sensitivity drives their widespread use, rapid response, adaptability to specific applications, and durability [49]. As a result, these nanocomposites' specifications and testing methods vary based on their intended uses [50]. Technological advancements, such as integrating dual ceramic components into an aluminum matrix to create hybrid composites, have facilitated the development of these models [51, 52]. Hybrid composites utilize two reinforcing materials to boost mechanical strength and reduce costs [53]. Additionally, the weight of the composite can be minimized to meet specific requirements [54]. Stir-casting techniques with mechanical stirring enable the even distribution of reinforcing components throughout the molten matrix [55, 56]. The hardness of the composite materials is assessed by applying stress, and performance metrics like fracture resistance and permanent deformation are evaluated according to standards set by the American Society for Testing and Materials (ASTM). For example, the hardness and tensile strength of steel are roughly proportional.

2.2. Different methods for the fabrication of composites

Various fabrication methods are employed to create these composites, each with distinct advantages and limitations. This discussion provides an overview of several key methods: liquid metallurgy (stir casting), squeeze casting, powder metallurgy, spark plasma sintering, friction stir processing, accumulative roll bonding, and ultrasonic-assisted stir casting.

Liquid metallurgy (stir casting): Liquid metallurgy, particularly stir casting, offers several advantages for producing metal matrix alloys. This method accommodates various shapes and sizes, including significant components weighing up to 5 kg. It can manage up to a 0.3 volume fraction of reinforcements with minimal damage to the materials. Economically, stir casting is the most cost-effective among various manufacturing techniques, providing excellent matrix-reinforcement bonding. Its versatility allows for the production of diverse shapes and sizes, making it a highly adaptable choice. Stir casting's ability to handle large volumes and its efficiency in maintaining material integrity while ensuring strong bonding between matrix and reinforcements make it an optimal method for producing high-quality metal matrix composites. The schematic illustration of stir casting is shown in Figure 2.1 [57–60].

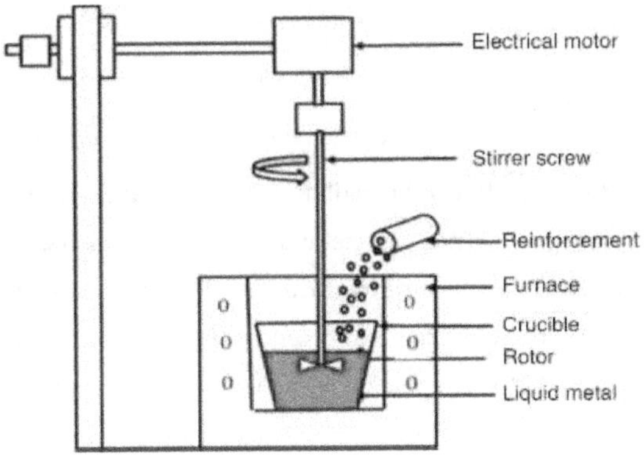

Figure 2.1. Schematic illustration of stir casting.

Squeeze casting: Squeeze casting is limited by mold shape, restricting it to parts no taller than 2 cm. It can accommodate up to 0.45 volume fraction of reinforcements. Despite its capacity, this method often causes significant damage to reinforcement materials, which can compromise the overall quality and performance of the composite. However, squeeze casting is recognized for its ability to produce high-quality components with excellent surface finish and minimal porosity. This method achieves a smooth surface and low air inclusion, enhancing the final product's visual and functional properties. Although it delivers superior results to some methods, squeeze casting is more expensive than stir casting. It is constrained by the mold's size limitations, making it unsuitable for more significant components. The schematic illustration of squeeze casting is shown in Figure 2.2 [50, 51, 61, 62].

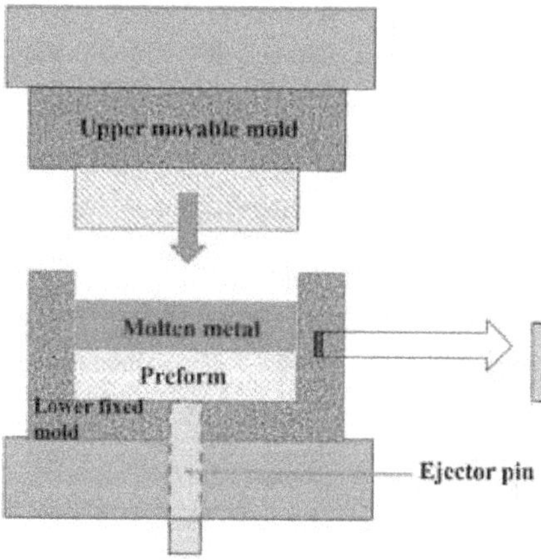

Figure 2.2. Schematic illustration of squeeze casting.

Powder metallurgy: Powder metallurgy offers versatility in producing a wide range of shapes, though it is limited in size and suitable for smaller components. The specific volume fraction of reinforcements it can handle is not well-defined, and one significant drawback is the potential for reinforcement fracture during processing. This method's primary advantage lies in producing parts that approach their final shape, reducing the need for additional machining. This near-net shape capability is beneficial for efficiency and material conservation. However, powder metallurgy is generally more costly than stir casting. Issues such as powder agglomeration can also affect the alloy's uniformity and consistency. Despite its advantages in shaping, the high cost and the risk of damaging reinforcements make powder metallurgy less favorable than other methods, particularly considering the overall performance and economic feasibility. The schematic illustration of powder metallurgy is shown in Figure 2.3 [63, 64].

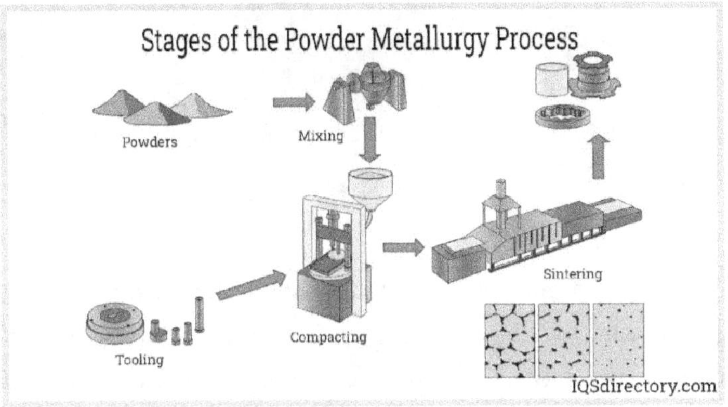

Figure 2.3. Schematic illustration of powder metallurgy.

Spark plasma sintering (SPS): SPS is a versatile technique capable of producing a variety of forms, though its size capabilities are somewhat limited. The specific volume fraction of reinforcements that can be incorporated using SPS is not specified. This method is precious for enhancing the interface bonding between the matrix and reinforcing materials, which it achieves effectively without causing significant damage to the reinforcements. Despite its benefits, the primary limitation of SPS is its high cost, which can restrict its applicability in specific scenarios. While SPS excels in improving material bonding and maintaining the integrity of reinforcements, the substantial expense associated with this process poses a challenge, making it less accessible for some applications. Consequently, despite its superior bonding capabilities, the cost factor remains a significant consideration when evaluating SPS for large-scale or cost-sensitive projects. The schematic illustration of spark plasma sintering is shown in Figure 2.4 [65–67].

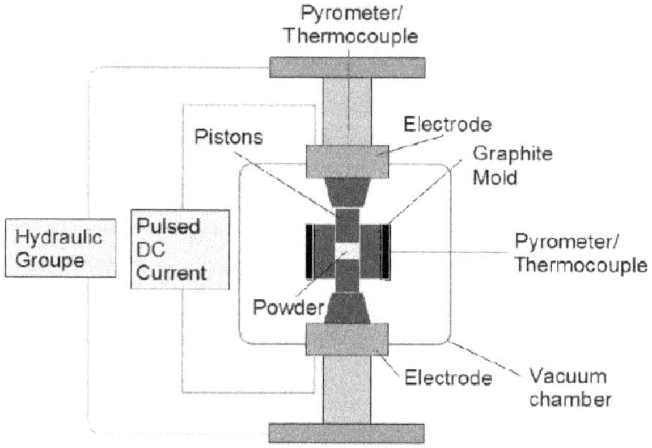

Figure 2.4. Schematic illustration of spark plasma sintering.

Friction stir processing (FSP): FSP is a versatile method capable of handling a wide range of forms, though it has limitations concerning the sizes it can process. The specific volume fraction of reinforcements used in FSP is not detailed. This technique is notable for enhancing surface properties and tensile strength without significantly damaging the reinforcements. However, FSP is relatively expensive and is predominantly utilized for surface modifications rather than large-scale production. This limitation in scope makes FSP more suited for applications where surface enhancement is crucial rather than for bulk manufacturing. Despite its effectiveness in improving material characteristics and maintaining reinforcement integrity, the high cost and focus on surface treatments restrict its broader application. Therefore, while FSP excels in specific areas, its cost and application limitations make it less feasible for extensive production needs. The schematic illustration of FSP is shown in Figure 2.5 [68–72].

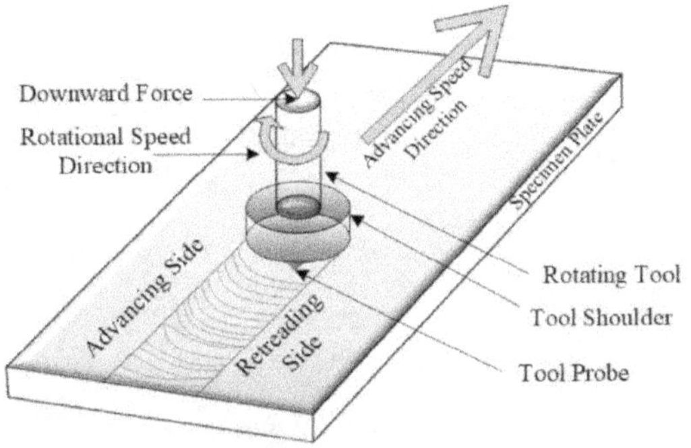

Figure 2.5. Schematic illustration of FSP.

Accumulative roll bonding (ARB): ARB is a versatile technique that can create a wide range of shapes, but it is limited in terms of the sizes it can handle. The specific volume fraction of reinforcements used in ARB is not specified. This method is advantageous because it minimizes damage to the reinforcements and effectively reduces porosity, resulting in stronger bonding between the matrix and reinforcements. Despite these benefits, ARB is relatively expensive and has limitations in size, which can restrict its practicality compared to more cost-effective methods. The higher cost and size constraints may make ARB less favorable for applications where budget and scale are critical considerations. While it excels in improving bond strength and reducing gaps, the financial and size limitations may impact its overall utility compared to more affordable and versatile alternatives. The schematic illustration of ARB is shown in Figure 2.6 [73–78].

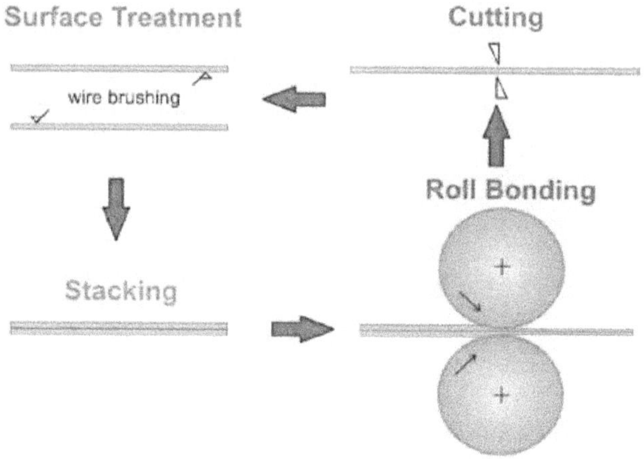

Figure 2.6. Schematic illustration of ARB.

Ultrasonic-assisted stir casting: Ultrasonic-assisted stir casting is a versatile method capable of producing MMCs with various shapes and sizes, including components up to 5 kg. This technique can accommodate up to 0.2 volume fraction of reinforcements, ensuring effective integration without causing significant damage to the materials. A key advantage of ultrasonic-assisted stir casting is its ability to enhance wettability and achieve a uniform distribution of nanoparticles throughout the matrix. This results in improved mechanical properties, such as increased strength and durability of the composites. Although it is moderately more expensive than traditional stir casting, its enhanced performance and quality justify the cost. The schematic illustration of ultrasonic-assisted stir casting is shown in Figure 2.7 [45, 79, 80].

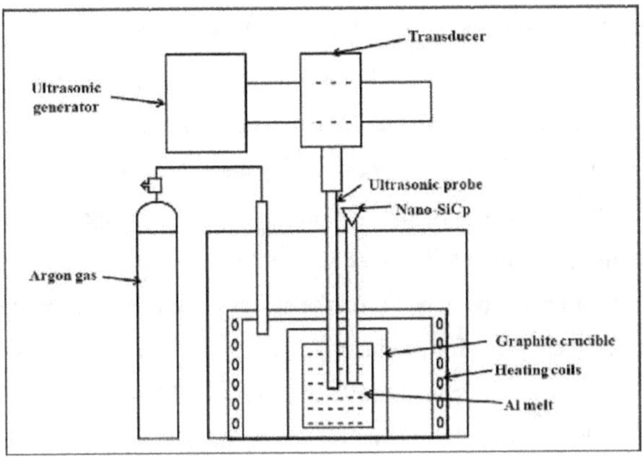

Figure 2.7. Schematic illustration of ultrasonic-assisted stir casting.

Compared to other manufacturing methods, ultrasonic-assisted stir casting is the most effective for producing high-quality MMCs. It outperforms alternative techniques regarding nanoparticle dispersion and wettability, which are crucial for achieving superior mechanical properties. Other methods, like squeeze casting and powder metallurgy, offer notable benefits, such as an excellent surface finish and the ability to produce near-net-shaped components. However, these methods also have significant limitations. Squeeze casting, for example, is restricted by mold shape and can cause considerable damage to reinforcements, leading to compromised composite properties. While effective for near-net shaping, powder metallurgy has high costs and potential reinforcement fracture issues. Ultrasonic-assisted stir casting strikes a favorable balance between cost and performance. Its ability to uniformly disperse nanoparticles and improve wettability makes it a preferred choice for advanced MMC production. As industrial applications increasingly demand high-performance materials with enhanced properties, ultrasonic-assisted stir casting is well-positioned to meet these requirements effectively. This method's versatility, performance enhancement, and cost-effectiveness ensure its continued relevance and utility in producing high-quality metal matrix composites.

2.3. Literature on the production of composites

A crucial component of the stir-casting process is the mechanical stirring within the furnace, which ensures that reinforcing materials are evenly distributed throughout the molten matrix, resulting in uniform composite properties across the entire material [81]. Various casting methods, including sand casting, die casting, and permanent mold casting, can be employed during stir-casting to handle the liquid alloy and matrix composites [82]. The choice of technique depends on the specific requirements of the outcome. Stir casting is particularly effective for producing composites with up to thirty percent reinforcement by volume, as this percentage allows for a well-balanced distribution of reinforcing particles, enhancing the composite's mechanical properties while maintaining manageable processing conditions [83]. Kocaman et al. [84] studied the effects of the ball milling (BM) process and different amounts of carbon nanotubes (CNTs) on the thermal and mechanical properties of hybrid reinforced composites. These composites were meticulously prepared by incorporating glass fiber (GF), novolac resin (No), and CNTs in various weight percentages (1%, 2%, and 4%). Different milling durations were used to improve the dispersion and aggregation of CNTs within the polymer matrix. The research found that a nanocomposite subjected to two hours of ball milling with 2% CNT reinforcement significantly enhanced all key parameters. Additionally, numerous studies have explored hybrid composites like Al-Mg-Si-SiC/muscovite/T6 [85], Al-SiC [86], and Al-PKSA-SiC [87, 88]. However, these composites often faced issues of reinforcement aggregation due to the absence of coatings on the reinforcing particles, and the extensive testing involved resulted in relatively high material and method costs.

Numerous recent studies have focused on hybrid composite materials such as Al-Mg-Si-SiC/muscovite/T6, Al-SiC, and Al-PKSA-SiC. However, these investigations revealed issues with reinforcement aggregation due to the removal of coatings on the reinforcing particles. The extensive testing required has led to high material and processing costs. Reinforced composites are increasingly preferred over traditional materials because they offer advantages like reduced weight, higher hardness, greater specific strength, and improved wear resistance. Nonetheless, optimizing the proportions of different materials requires considerable time and material investment. This study aims to develop methods for selecting material mixtures to enhance bonding strength between the matrix and reinforcements, minimize aggregation, and reduce costs. The hybrid composites under development will be Al 6061-SiC-B_4C-GO, which are expected to lower manufacturing costs and simplify structural design. These composites will be evaluated for wear resistance, tensile strength, and hardness. The addition of silicon carbide is anticipated to improve tensile strength. Graphene oxide, with its unique monomolecular functional groups (hydroxyl, carboxyl, epoxide, and carbonyl), enhances electrochemical, thermal, and chemical properties. The study will also employ the strawberry fitness function to optimize the tribological properties of the composites, aiming to establish a predictive system for enhancing tribological performance.

In addition to studying Aluminum-Yttrium Oxide [87], Zinc Oxide Nanorods [88], and Graphene Oxide Reinforced Aluminum Matrix Composites [89], scanning electron microscopy (SEM) coupled with energy-dispersive spectroscopy (EDS) was utilized to analyze the microstructural changes and worn surface characteristics of the fabricated samples [90]. The consolidation of hybrid composite powders was achieved through a hot-pressing process at 515°C under a pressure of 700 MPa for 3 hours [91]. Previous attempts to identify the optimal metal matrix composites failed to find the ideal solution or the most efficient mechanism for achieving this. Consequently, several mixtures yielded suboptimal results. To address this, the current study employed an optimized mix selection technique, which does not require physical testing to determine the best mixtures. Instead, MATLAB programming was used to determine the optimal selection intelligently. The optimal range of test performance scores—including flexural strength, impact strength, tensile strength, hardness, wear rate, and coefficient of friction (COF)—was updated using the novel Strawberry Optimization Method for Neural Networks (SRNMS). The classification layer of SRNMS was refined with strawberry optimization techniques, which adhere to standards set by the ASTM. ASTM's standards are used for tensile testing to measure the force needed to break a polymer composite and the elongation before breakage. This helps develop stress-strain diagrams and compute tensile modulus. ASTM also aids in evaluating physical properties such as shear, flexural, tensile, and compressive strengths of composite materials used in structural applications. The mix selection procedure was repeated until the predetermined test range was met. If any selected mix matches the saved test results range, the procedure concludes, and the performance is verified. Recent discoveries in hybrid composites have resulted in significant material science and engineering advancements. Some of the most notable works are mentioned in this field below:

In recent years, aluminum alloys have become essential across various industries due to their extensive range of properties and versatility. Consequently, they are widely used in manufacturing parts for automobiles, watercraft, aircraft, and submarines. A notable study by Sharma et al. [85] focused on the fabrication of aluminum composites reinforced with silicon carbide and hydrated aluminum-potassium-silicate/muscovite/aluminosilicate (Al-Mg-Si-SiC/muscovite/T6). These composites were developed explicitly for high-strength applications. The study revealed that incorporating muscovite or aluminosilicate significantly enhances the softening properties of matrix composites, making them suitable for applications requiring high flexibility. However, the research also pointed out that the lack of proper coating on the reinforcements could lead to particle aggregation, which might compromise the uniformity and overall performance of the composite material. This emphasizes the importance of addressing reinforcement coating issues to fully optimize the properties and performance of aluminum-based composites in advanced engineering applications.

Sachin et al. [86] employed the stir casting process to extensively examine Al 6061 reinforced with SiC. They varied the weight percentages of SiC within the alloy and assessed the material's mechanical properties, including hardness, wear resistance, and tensile strength. The test samples were meticulously prepared according to ASTM standards to ensure the accuracy and reliability of the results. The study found that the addition of SiC significantly enhanced the mechanical properties of the Al 6061 base material, improving both hardness and wear resistance, which contributed to the composite's overall durability. Moreover, it was observed that increasing the SiC weight percentage in Al 6061 led to a greater sliding distance, indicating that while the reinforcement improves specific mechanical properties, it can also influence other aspects, such as sliding wear characteristics. The results underscore the importance of carefully adjusting the SiC weight percentage to achieve a balanced set of mechanical properties suitable for specific applications.

Sharma et al. [85] investigated Al-Mg-Si-SiC/muscovite/T6 composites produced through stir casting. The study revealed that adding muscovite and aluminosilicate to the aluminum matrix resulted in a softer matrix and enhanced the overall strength of the composites. However, a fundamental limitation was that the bonding strength between the matrix and the reinforcing materials did not significantly improve. This issue was attributed to the lack of coating on the reinforcing materials, which hindered their complete integration with the matrix. As a result, while the reinforcements improved the composite's strength, the uncoated reinforcements limited the potential gains in bond strength.

Ikubanni et al. [87] explored the potential of fabricating aluminum composites reinforced with SiC and palm kernel shell ash (PKSA) using the stir casting method. Their research focused on evaluating the mechanical properties of the Al-PKSA-SiC composites, particularly their density and strength. The study revealed that integrating PKSA and SiC into the aluminum matrix significantly enhanced the composites' density and strength. PKSA, derived from agricultural waste, was an effective and eco-friendly reinforcing material, contributing positively to the composite's overall mechanical performance. However, the study also identified challenges in specimen collection and processing. Collecting palm kernel shells, converting them into ash, and preparing the test specimens were considered time-consuming. This added time might be seen as a drawback when considering the scalability and practicality of these composites for industrial applications. Despite these challenges, the research highlights the feasibility of creating high-performance aluminum composites by combining SiC with agricultural by-products like PKSA.

Sharma and Meena [92] explored hybrid aluminum composites reinforced with titanium dioxide (TiO_2) using the stir-casting method. Their research primarily aimed to assess the mechanical properties of the Al-TiO_2 composites. The findings indicated that adding TiO_2 to the aluminum matrix significantly enhanced its mechanical properties, including hardness and tensile strength. The uniform distribution of TiO_2 particles throughout the matrix improved the composite's load-bearing capacity and overall structural integrity. However, the study was limited in scope, focusing mainly on basic mechanical properties and omitting critical performance aspects such as abrasion resistance and wear behavior. These factors are crucial for evaluating the suitability of hybrid composites in practical applications where friction and wear are prevalent. Despite these limitations, the research underscores TiO_2's potential as a valuable reinforcing material for aluminum-based hybrid composites.

Using the stir-casting method, Satishkumar et al. [93] examined aluminum composites reinforced with SiC and fly ash (FA). The study aimed to assess how these reinforcements affect the mechanical properties of the composite material. The results showed that adding SiC and FA significantly enhanced the hardness and tensile strength of the aluminum matrix, indicating improved mechanical performance. This enhancement is likely due to SiC's hard and rigid nature and FA's lightweight yet sturdy characteristics, leading to a more durable and robust composite. However, the study also identified a notable drawback: exceeding a certain reinforcement threshold could negatively impact the composite's overall strength. Excessive reinforcement might cause issues such as particle clumping and inadequate interfacial bonding, leading to weak spots in the composite and potentially compromising its mechanical performance.

Employing the stir casting method, Anand et al. [94] conducted a detailed study on aluminum MMCs that incorporated nano-alumina (Al_2O_3) and molybdenum disulfide (MoS_2) within an Al 6061 matrix. Their research evaluated the resulting composites' impact strength and wear rate. The addition of nano-Al_2O_3 and MoS_2 significantly increased the impact strength of the Al 6061 matrix while reducing the wear rate. This indicates that these reinforcements enhance the composite's durability and resilience under dynamic loading conditions. Additionally, the study found that incorporating nano-Al_2O_3 and MoS_2 also improved the hardness of the MMCs. This hardness improvement is attributed to the high strength of nano-Al_2O_3 particles combined with the lubricating properties of MoS_2, resulting in a more wear-resistant and durable composite material. However, despite these significant enhancements, the study does not address critical issues such as the distribution of nano-sized reinforcements and their impact on the overall processability and cost of the composites.

Pagidi Madhukar et al. [79] employed the stir-casting technique to investigate hybrid aluminum composites. These composites were created using Al 7150 reinforced with hexagonal boron nitride (hBN). Their study aimed to evaluate the mechanical strength of these composites. The results demonstrated that the hybrid Al 7150-hBN composites exhibited improved overall mechanical performance. Specifically, including hBN enhanced the composite's heat conductivity and lubrication properties, positively impacting its performance. However, the study also identified a drawback: adding certain chemicals or additional reinforcements reduced the composites' tensile strength. This decrease in tensile strength may be due to the formation of weak interfaces within the composite or insufficient bonding between the matrix and the reinforcement.

Satishkumar et al. [93] conducted experiments on hybrid composites made from Al 6061 reinforced with SiC and fly ash (FA), referred to as SiC-FA-Al. The stir-casting technique was employed during manufacturing to ensure uniform distribution of the reinforcements within the composites, involving the agitation of the molten aluminum matrix. The study focused on evaluating the mechanical properties of these composites, including tensile strength, hardness, and wear characteristics. The results indicated that adding SiC and FA to the Al 6061 matrix enhanced its tensile strength and hardness. However, it was also noted that excessive reinforcement could negatively impact the strength of the composite specimens. The detailed mechanical properties of the hybrid composites are summarized in Table 2.1, which is available in the study.

Table 2.1. Mechanical properties of hybrid composites.

S. No	**Mechanical properties/ Composite materials**	**Tensile strength (MPa)**	**Density (g/m^3)**	**Elastic modulus (GPa)**	**Hardness Vickers**
1	Al	125	2.660	70.3	200
2	Mg	260	1.74	44.80	48.3
3	SiC	250	>3.1	410	>93
4	PKSA	281.71	1.725	40.2	98
5	TiO_2	33.07	1.29	3.17	330
6	H	845	1.46	210	38-54
7	FA	9.06	3.923	353.6	90

8	Nano Al_2O_3	342	3.97	28.90	1700-2000
9	Al7150	160	2.83	71	140
10	hBN	2.41	2.21	64.8	19-192

2.4. Contact stress analysis

Aluminum composite gears are increasingly used in various industrial applications due to their lightweight, high strength, and excellent wear resistance. However, one of the main challenges in designing and analyzing these gears is the accurate measurement of contact stress, a critical factor affecting their performance and durability. In mechanical engineering, contact stress is a critical parameter, especially in the design and functionality of gears. Within a gear system, it refers to the force per unit area applied to the surfaces that come into contact with each other. Understanding this concept is crucial for analyzing how gears interact under load and is a key factor in assessing their durability and operational efficiency. Contact stress arises at the points where gear teeth mesh during operation. It is generated by the forces transmitted through the gear teeth as they engage and transfer torque between gears. Evaluating the magnitude and distribution of this contact stress is essential for determining the performance and longevity of gears [23, 95, 96].

One of the significant issues related to contact stress is its impact on surface degradation. High contact stresses can lead to various forms of surface deterioration, including pitting, spalling, and scuffing. For instance, pitting happens when specific areas of a gear surface are subjected to repeated stress cycles, causing material displacement and the formation of small pits or holes. Over time, this can significantly reduce the gear's efficiency and performance. On the other hand, scuffing results from severe surface wear due to insufficient lubrication and high friction, leading to metal-to-metal contact and rapid degradation of the gear surfaces. Scuffing is usually evident upon examining the gear surfaces. Contact stress also plays a crucial role in wear rates; the level of contact stress directly influences how quickly gears wear out. High contact stress can accelerate wear, increasing maintenance needs and the risk of gear failure. While proper lubrication and surface treatments can mitigate wear, they cannot eliminate the effects of high-contact stress. Thus, understanding and managing contact stress is vital for extending gear lifespan and reducing maintenance costs. Additionally, contact stress significantly affects fatigue. Repeated exposure to high-contact stresses can induce fatigue, leading to cracks and gear failure. Fatigue failure is often marked by surface fractures that spread over time, undermining the gear's structural integrity. It is crucial to consider fatigue resistance during gear design and material selection to ensure reliable performance under various operating conditions [24, 97, 98].

Regarding gear system performance and longevity, contact stress is a crucial factor to consider. It refers to the force exerted per unit area between two surfaces in contact, specifically the teeth of gears. Misalignment is a primary factor that can lead to increased contact stress, significantly affecting the functionality and lifespan of gear systems. Understanding various misalignments—such as angular, axial, and radial—is essential for identifying issues and applying effective solutions to prevent early wear, noise, and failure [29, 29, 99].

Angular misalignment occurs when the axes of two mating gears are not parallel and intersect at an angle. This misalignment causes the gear teeth to engage at an angle not perpendicular to the rotation axis. As a result, the contact stress is unevenly distributed along the teeth, leading to several adverse effects. The skewed contact angle creates localized high-stress areas that accelerate wear and may lead to pitting and surface fatigue. Additionally, the uneven engagement of the teeth generates noise and vibrations, which signal potential mechanical failure and contribute to an unpleasant operating environment. The efficiency of the gear system is also compromised, as increased friction and energy loss reduce overall transmission performance.

Axial misalignment occurs when gears are incorrectly positioned along the axial direction, leading to an offset between the gear teeth. This misalignment can have several adverse effects. Initially, it increases the axial forces exerted on the bearings and shafts, potentially causing higher bearing loads, shaft bending, and accelerated wear on the components. Additionally, the offset results in unequal tooth engagement, leading to uneven contact stress on specific teeth. This can cause localized wear and reduce the gear's lifespan. Axial misalignment also negatively affects the overall performance of the gear system by decreasing operational smoothness and impairing torque transfer, ultimately impacting the machinery's reliability and efficiency.

Radial misalignment occurs when the axes of two gears are not perfectly parallel. This deviation from parallelism leads to gaps or interference between the gear teeth, causing several significant issues. Firstly, it affects the engagement of the gear teeth, where gaps can prevent full engagement, and severe interference can cause teeth to clash, resulting in increased stress and potential damage. Additionally, the irregular spacing between teeth due to radial misalignment heightens contact stress at specific points, potentially leading to gear tooth failure, such as chipping or cracking. This misalignment also increases friction and energy losses within the gear system, which can cause overheating and shorten the overall lifespan of the gears. As a result, radial misalignment introduces operational inefficiencies. Different types of gear misalignments are shown in Figure 2.8 [2, 5, 8, 13, 18].

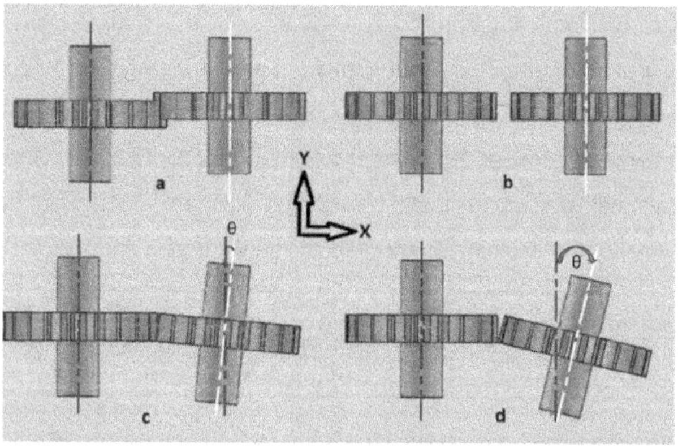

Figure 2.8. Representation of the gear pair alignment: a) Axial, b) Radial, c) Pitch and d) Yaw

The misalignment of gear systems has a significant impact on both their performance and longevity. One of the significant consequences is accelerated wear and tear, which occurs when misalignment increases contact stress, causing excessive wear on gear teeth. This rapid wear leads to a loss of tooth profile accuracy, decreased mating efficiency, and eventual gear failure. Misalignment also shortens gear life by creating irregular contact patterns and increased stress concentrations, undermining the gears' fatigue resistance. As the accumulated damage from misalignment progresses, the operational lifespan of the gear system is reduced, leading to diminished productivity. Furthermore, misalignment results in higher maintenance costs, as frequent adjustments, replacements, and repairs are necessary to address the persistent issues. These increased maintenance requirements lead to higher operating costs, making misalignment an expensive problem affecting gear systems' reliability and economic efficiency [8].

Gears are essential for transferring torque and motion between shafts through interlocking teeth in mechanical systems. Historically, gears have been made from various materials, but aluminum composites have gained popularity due to their superior properties. Compared to polymer alloys, aluminum composites offer advantages such as exceptional strength, excellent shape stability, and high corrosion resistance. These qualities make them highly effective. Aluminum's malleability allows for precise dimensional control and complex cutting processes, making it ideal for manufacturing intricate gear designs. Additionally, aluminum's ductility and non-magnetic properties enhance its suitability for gear production. A key benefit of aluminum is its ability to be shaped into various forms while retaining its metal properties, resulting in gears with increased strength and durability. Aluminum gears are known for their smooth operation and high efficiency, as they minimize wear and extend the operational life of gear systems. These attributes make aluminum an attractive choice for gear manufacturing, combining high performance with reliability in demanding applications [7, 31].

2.5. Literature studies on contact stresses

Limited studies focused on the contact stress on composite gears. Catera et al. [100] introduced a metal with composite gear. They compared a spur hybrid gear to a less-weight steel gear of similar mass to investigate the impacts of the composite web on the stiffness performance. It attains better quasi-isotropic and macro-geometry behaviors and low transmission error.

Maoa et al. [9] investigated the tribological properties of acetal and nylon polymer gears and explored the effect of load, geometry, and surface temperature along with the driver and driven influence. They employed a unique test method to measure wear rate, explicitly identifying the different failure modes of nylon gears compared to acetal gears. They aimed to establish a relation between gear surface temperature and loading capacity for acetal gears. They finally observed the substantial variance in tribological properties when operating acetal against nylon gears.

Miryam et al. [28] developed a new method for calculating contact stress in spur and helical gears with a transverse contact ratio of more than two. The method combines the Hertz equation with an improved load dispersion model based on the minimal elastic potential measure. They aimed to determine critical loading and contact stress values and recommend pit loading capability measures. They finally enhanced the design and durability of high transverse contact ratio gears under non-standard conditions while advancing the understanding of load distribution, contact stress, and gear failure.

Wennian et al. [27] introduced an analytical methodology to investigate the corner contact influences of spur gear transmission systems with profile modifications. The study highlighted the importance of considering corner contact influences in gear dynamic analysis by differentiating 2 kinds of gear mesh stiffness models and generalizing 3 generally employed single DOF models.

Hotait et al. [5, 14] presented a unique set of investigated results, including dynamic transmission error and stress factor measurements obtained concurrently from the same spur gears. The study aimed to connect gear durability research with noise and vibration by establishing the relation between dynamic transmission error and stress factor. The study also concluded that the dynamic test rig, integrating root strain and dynamic transmission error measuring system, presented dynamic transmission error and stress factor measurements from un-modified and modified gears.

Huseyin et al. [21] investigated the impact of modifying the width of teeth on Nylon 6 spur gears, which are prone to fracture owing to heat concentration in the single-tooth mesh region. The objective is to enhance load distribution in the mating areas by broadening the single-tooth region. The experimental evaluation of changed and unchanged gears under 3 loading conditions includes measuring the tooth temperature and wear rate. The study aims to determine whether modifying teeth width can reduce the wear rate of Nylon 6 gears and enhance their performance. The findings can provide valuable insights for enhancing the durability and reliability of Nylon 6 plastic gears.

Senthilvelan et al. [101] investigated the wear behavior of gears made by the polymer composite employing a power absorption-kind gear setup. Nylon-6 and glass-filled Nylon-6 (GFN6) gears have been injection-molded and underwent different speed conditions and load values. The study examined the impact of revolution speed on tooth wear of unfilled and revolution GFN6 and the impact of revolution lag owing to differences in the elastic module of mating materials. GFN6 gears exhibited superior wear resistance owing to the enhanced compression strength and modulus of elasticity. The study also analyzed wear particles to determine the wear mode influenced by contacting stresses and the gear material's hardness.

Along with the experimental analysis, various authors conducted simulations to investigate the contact stresses on composite gears. Shutting et al. [18] employed FEA to examine the influence of centrifugal loads on tooth root bending stresses of spur gears. The study aims to incorporate this effect in the bending strength analysis of the gears at high speed. The study also conducted loaded tooth contact analyses to calculate gear contact and root bending stresses when engaged with a solid mating gear under a torque load. Additionally, the study compared the effect of centrifugal deformation on the tooth and root bending stresses of the inclined web gears with that of straight web gears. It explained why the former is not affected by centrifugal deformation.

Sankar et al. [12] investigated tooth failure in spur gears. They proposed a design modification to prevent damage to gear teeth by using circular root fillets instead of unique trochoidal root fillets. Gear manufacturing can weaken gears with fewer than 17 teeth through undercutting at the root. The strength of the modified teeth with circular root fillets is compared to the standard design employing the ANSYS tool, and the bending strength is analyzed. Results show that the novel design with circular root fillets has greater bending strength than the unique trochoidal root fillet gear. The study suggests using circular root fillets for pinions with fewer teeth and trochoidal root fillet gear for gears with more teeth.

Shao et al. [102] proposed an improved calculation method for friction innervation in spiral gears having mesh misalignment and sliding friction to better understand changes in their dynamic characteristics. Due to errors in manufacturing and assembling, mesh misalignment is inevitable and can alter the dynamic characteristics of spiral gears. The suggested methodology used the time-varying length of the contact line and friction coefficient model to determine the time-varying sliding frictional force. The analytical model was developed, and simulations of different COF models and mesh misalignments were conducted. This enables a novel methodology for analyzing the excitation characteristics of spiral gears having mesh misalignment.

2.6. Relevant literature on the present study

Mishra et al. [56] emphasize that in recent years, aluminum alloys and composites have increasingly been chosen for fabricating aluminum matrix composites. This trend is likely due to their superior material properties and broad range of applications. Their study focuses on using stir casting to create Al 6061 as the matrix material, with SiC and Al_2O_3 as reinforcements. The research investigates the manufacturing process, its parameters, and the impact of these reinforcements on the mechanical properties of aluminum metal matrix composites (Al MMCs). Findings suggest that stir casting is preferred for producing aluminum-based hybrid MMCs and that increasing reinforcement percentages enhances tribological and mechanical properties. Hybrid composites, in particular, show superior material qualities compared to those with a single reinforcement.

Venkatesh et al. [103] investigated the synthesis of aluminum composites using an ultrasonically assisted stir-casting process. They experimented with various weight percentages of SiC nanoparticles—0.5%, 1.0%, 1.5%, and 2.0%. Ultrasonication refined the grain size and ensured uniform dispersion of nano-SiC throughout the aluminum matrix. SEM confirmed that nano-SiC particles were evenly distributed in composites with 0.5%, 1.0%, and 1.5% SiC. However, X-ray diffraction (XRD) revealed the presence of Al_2Cu intermetallic phases in the composite with 2.0% SiC. Adding SiC reinforcements significantly enhanced the mechanical properties of the composites. The composite with 1.5% SiC achieved a maximum hardness of 163 BHN and an ultimate tensile strength (UTS) of 431 MPa. In contrast, increasing the SiC content to 2.0% reduced mechanical properties due to brittle agglomerations caused by thermal expansion mismatches between the matrix and the SiC particles. Various strengthening mechanisms were explored, revealing that thermal mismatch had the most significant impact, followed by the Orowan mechanism and grain refinement. The study found that adding SiC nanoparticles increased the effectiveness of these strengthening processes. Fractographic analysis showed brittle fracture patterns in SiC-reinforced composites and ductile fracture in unreinforced aluminum, with the SiC-reinforced composites exhibiting cleavage patterns indicative of brittle fracture.

Vibhu Singh et al. [104] investigated the production of an aluminum metal matrix composite using Al 6061-T6 alloy as the matrix material. The composite included 2% by weight of hard ceramic titanium diboride (TiB_2, with a particle size of 14 μm) and 0.5% by weight of rare earth oxide lanthanum oxide (La_2O_3, with a particle size of 40–50 nm) as reinforcing powders. The composite was manufactured through ultrasonic-assisted stir casting. The study examined the composite's microstructure, residual stress, and mechanical properties and conducted pin-on-disk wear tests. The wear rate was optimized by varying load, sliding distance, and sliding speed, using Taguchi's optimization and analysis of variance. Ultrasonic stirring led to a uniform distribution of TiB_2 and La_2O_3 particles in the Al 6061-T6 matrix, as confirmed by optical micrography and field emission scanning electron microscopy (FESEM) with energy-dispersive spectrometry (EDS). The composite exhibited a 25.15% increase in microhardness and a 31.32% improvement in tensile strength compared to the base alloy, attributed to a higher dislocation density. Analysis of variance showed that load was the most significant factor affecting wear rate, contributing 69.54% to the total variation, followed by sliding distance (17.45%) and sliding speed (12.56%). The optimal parameters for minimizing wear rate were identified as a load of 20 N, a sliding distance of 1500 meters, and a sliding speed of 2.5 meters per second. SEM analysis of worn surfaces revealed various wear mechanisms, including abrasion, delamination, and adhesion, with abrasion being the most dominant. Incorporating TiB_2 and La_2O_3 particles increased residual stress, shifting from tensile to compressive stress from the inner to the outer surface of the composite casting.

Ahmad Ghahremani et al. [105] investigated metal matrix nanocomposites (MMNCs) due to their outstanding mechanical, chemical, thermal, and tribological properties, making them suitable for high-reliability applications. While aluminum and magnesium alloys in extreme wear environments are well-studied, the wear behavior of nanocomposites remains less understood. This review examines MMNC surface characteristics, focusing on manufacturing processes, nanoparticles, and dominant wear mechanisms. Key findings include: ceramics-reinforced composites generally offer better wear resistance than those with carbon-based nanomaterials; hybrid MMNCs with multiple reinforcements, particularly ceramic-carbon combinations, enhance surface resistance; solid-state processes like powder metallurgy usually provide superior wear resistance compared to liquid processing methods; smaller reinforcements often improve performance; and abrasion is the most prevalent wear mechanism. The study highlights existing research gaps and suggests further exploration in the field. They explored the effect of ultrasonic vibration on Mg/Al bimetals manufactured via lost foam compound casting (LFCC). Without ultrasonic vibration, the interface layers were Al_3Mg_2 + Mg_2Si, $Al_{12}Mg_{17}$ + Mg_2Si, and $Al_{12}Mg_{17}$ + δ-Mg eutectic. Ultrasonic vibration refined these phases, evenly distributed Mg_2Si, and eliminated oxidation defects. This resulted in a more uniform microhardness and an 86.5% increase in shear strength, reaching 69 MPa. The untreated bimetal showed a brittle fracture, while the ultrasonic-treated bimetal exhibited improved shear strength and partial plastic deformation at the interface due to enhanced microstructural refinement.

Bhavesh Chaudhary et al. [106] explored Friction Stir Powder Additive Manufacturing (FSPAM) as a solution to the challenges of Fusion-Based Additive Manufacturing (FBAM) for aluminum alloys and their metal matrix composites (MMCs). These challenges include high thermal conductivity, porosity, cracks, thermal distortion, and unwanted phase formation. The study focused on multi-layer depositions of Al 6061-based MMCs, specifically Al 6061/6wt.%FeCoNi and Al 6061/6wt.%Ni. Key areas investigated were microstructure, particle distribution, phase analysis, residual stresses, and their effects on microhardness, tensile strength, and wear characteristics. The researchers minimized material buildup on the advancing side by adjusting tool rotation direction in successive layers, achieving smoother surfaces. FSPAM-produced MMCs displayed uniform particle distribution, strong interlayer bonding without defects, refined equiaxed grains due to dynamic recrystallization, and 40 and 50 MPa compressive residual stresses. No harmful intermetallic compounds formed, and a bowl-shaped interface was achieved. Compared to Al 6061 alloy depositions, the MMCs showed increased microhardness, ultimate tensile strength, and improved wear resistance despite reduced elongation. The study confirms FSPAM as a viable alternative to FBAM for high-quality aluminum alloy-based MMCs.

John Victor Christy et al. [107] comprehensively review friction stir welding (FSW) techniques for aluminum alloys and metal matrix composites (Al MMCs), highlighting their importance in the construction, aerospace, and automotive industries. They address challenges associated with the fusion welding of aluminum alloys, such as issues with non-heat-treatable alloys, porosity, and liquation fractures. Despite its widespread use, conventional in-air FSW has limitations in welding these materials effectively. The paper explores newer FSW methods, including underwater friction stir welding (UFSW) and vibrational friction stir welding (VFSW), which aim to overcome these limitations. VFSW is noted for its cost-effectiveness, energy efficiency, and versatility. The study also underscores the need for further research on the FSW of Al-MMCs to enhance joint quality and address existing research gaps.

Dinesh Kumar Koli et al. [108] investigated the impact of ultrasonic-assisted stir casting on the mechanical properties of Al 6061 aluminum alloy reinforced with nano-sized Al_2O_3 particles (40 nm, 1-4 wt.%). Their study focused on these nanoparticles' microstructural and mechanical effects within the aluminum matrix. The results showed that ultrasonic vibrations at 20 kHz effectively dispersed the nano-Al_2O_3 particles uniformly throughout the matrix, reducing particle clustering. This uniform distribution led to slight improvements in hardness, tensile strength, and compressive strength compared to conventional stir-casting methods. The enhanced distribution of nanoparticles contributed to better load transmission and structural integrity.

Pankaj Kumar Gupta et al. [109] reviewed the fabrication techniques for aluminum and its alloy metal matrix composites (MMCs), emphasizing ceramic powder reinforcement. The review highlights aluminum's growing importance in nuclear, aerospace, and automotive industries due to its favorable physical and mechanical properties. Stir casting is noted as the most effective method for its simplicity and cost-effectiveness, especially for mass production. The review explores enhancements in stir casting through processes like electromagnetic generation, ultrasonic aid, thermal treatment (T6), forming, and sintering, which improves grain structure and mechanical properties. The study underscores significant performance gains in aluminum-based MMCs achieved through these advanced processes.

Barada Prasanna Sahoo et al. [110] comprehensively analyzed liquid-state processing techniques for aluminum-based metal matrix nanocomposites (MMNCs), focusing on their microstructural properties. The study synthesizes findings from various investigations, highlighting that mechanical milling followed by stir casting is highly effective for producing nanocomposites with fine grain sizes, interfacial solid bonding, and uniform reinforcement dispersion with minimal accumulation. Ultrasonic-assisted casting further enhances these properties by reducing particle agglomeration and improving nanoparticle wettability. The study also covers advanced techniques like induction melting, which aids in achieving well-distributed reinforcements along grain boundaries, and electromagnetic casting, known for reducing cracks and cavity defects. Additionally, spray forming and disintegrated melt deposition are noted for producing near-net-shaped MMNCs with uniform nanoparticle distribution. Despite these advancements, the research identifies ongoing challenges and gaps in the liquid-state processing of aluminum MMNCs, suggesting that future studies should focus on addressing these issues to enhance processing methods and performance.

Raviteja Surakasi et al. [111] explore using nano-sized B_4C particles in aluminum MMCs to address new technical demands in industries like aerospace and automotive. Their research focuses on Al 6061 alloy composites reinforced with B_4C nanoparticles, ranging from 0% to 8%, using an ultrasonic-cavitation casting technique. This method effectively ensures a uniform dispersion of B_4C within the aluminum matrix. The study finds that increasing B_4C content improves the composite's hardness, ultimate tensile strength, and yield strength. However, beyond a specific concentration, the ultimate tensile strength declines. The research underscores the benefits of ultrasonic-cavitation casting in enhancing MMCs but also points out the need to balance reinforcement levels to maximize advantages while minimizing potential drawbacks.

In this study, Balamurugan et al. [112] investigate how adding titanium carbide (TiC) and rice husk ash (RHA) to Al 6061 hybrid composites can enhance wear and erosion resistance. The research utilizes a two-step ultrasonic-assisted stir casting method to create four different Al 6061 hybrid composites with varying TiC percentages (3%, 6%, 9%, and 12%) while keeping the RHA percentage constant at 3%. Solid particle erosion tests are conducted using a Taguchi L27 (3^3) orthogonal array to analyze the effects of different parameters—erodent particle impact angle, velocity, and discharge rate—on erosion rates. The results show that increasing TiC content up to 12% significantly improves erosion resistance due to the hardness and wear resistance of TiC particles. SEM analysis of surface failures reveals that higher TiC content reduces surface imperfections and enhances erosion resistance. The study concludes that incorporating TiC and RHA into Al 6061 composites significantly improves their durability, making them suitable for high-wear applications, including military use.

Sukanta Sarkar et al. [113] explored three dispersion methods—Direct Drop of Carbon Nanotubes (DD), Mixed Aluminum and CNT Pellet Drop (PD), and Direct Drop of Ultrasonicated CNT (USDD)—for creating Al 6061–0.5 wt.% CNT composites using combined stir-casting and squeeze-casting. They found that the PD method achieved the best CNT dispersion and retention, verified by transmission electron microscopy (TEM), selected area electron diffraction (SAED), and XRD. The PD composite significantly improved hardness, tensile strength (40% increase), and yield strength (33% increase) compared to as-cast Al 6061 while maintaining ductility. Fracture surface analysis revealed robust CNT bridging and ductile fracture, strengthening effective load transfer. The study concluded that the PD process is highly effective for producing Al 6061-CNT composites with enhanced mechanical properties and CNT dispersion.

2.7. Summary of the literature

- The literature review indicates several manufacturing techniques for producing MMCs reinforced with nanoparticles. Among liquid processes, Ultrasonic-Assisted-Stir casting has yielded particularly favorable outcomes.
- Hybrid nanocomposites, which incorporate low volumes of multiple nanoparticles, demonstrate superior mechanical and tribological properties compared to those reinforced with a single type of nanoparticle. Notably, Orowan and dislocation mismatch strengthening is more pronounced when particle sizes are reduced from 100 nm to 1 nm.
- Numerous researchers have reinforced aluminum with various fillers, enhancing its tribological and mechanical performance. This results in improved strength, wear resistance, and friction coefficients, expanding the potential applications of these composites across various industries.

2.8. Literature gaps

From the extensive relative literature of the present study, the following gaps were identified;

- Research on the synergistic effects of combining graphene oxide (GO), silicon carbide (SiC), and boron carbide (B_4C) as fillers in aluminum composites is minimal. Investigating these combinations could uncover enhanced mechanical and tribological properties, potentially leading to advanced composite materials with superior performance in various applications.
- There is a significant gap in experimental studies evaluating contact stresses and strains in gears under misalignment conditions. Such research is crucial to understanding gears' real-world performance and durability, providing insights into improving design and manufacturing processes to mitigate misalignment effects and enhance gear longevity and reliability.

2.9. Objectives of Proposed Research

Based on the literature gaps, the present work is divided into two objectives (or phases), and each phase has several sub-objectives;

Objective-1: Architecture for identification of the best mixture

- Initially, Al 6061-T6 is selected as a base material, and silicon carbide (SiC), boron carbide (B_4C), and Graphene oxide (GO) are selected as reinforcement materials.
- Furthermore, a novel Strawberry Recurrent Neural-based Mix Selection (SRNMS) was mathematically designed in MATLAB to save the material, time, and cost of this preparation. Based on the developed framework, the effective percentages of mixes are selected.
- The specimens were prepared using the stir-casting method based on the selected mixes. Then, the test samples are tested per ASTM standards to test the tribological and mechanical properties. Also, the worn surface analysis is carried out using SEM images. The block diagram of Phase-I is shown in Figure 2.9.

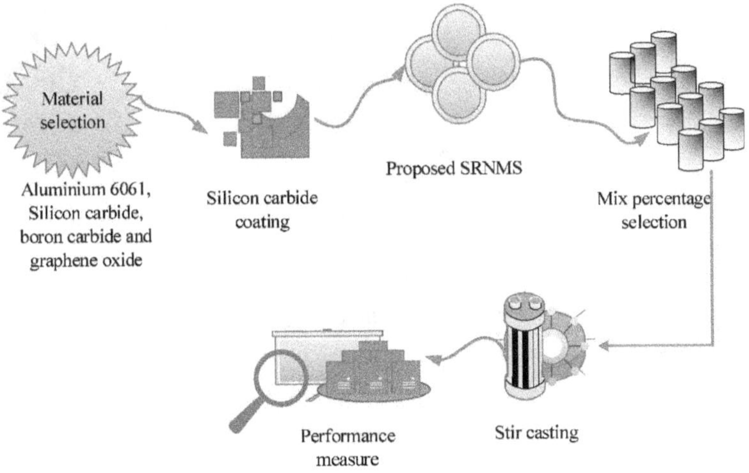

Figure 2.9. Block diagram of Phase-I work.

Objective 2: Performance evaluation of composite gears

- Graphene oxide (GO), SiC, and B_4C with Al 6061-T6 hybrid composite were considered for producing gears.
- The gear pair was fabricated using Wire EDM Cutting.
- The universal gear test rig was developed to test gears under various misalignment positions.
- The contact strains and stresses were analyzed using the developed rig and simulated using FEA.
- Finally, the results of the FEA and experiment were compared. The block diagram of Phase-II is shown in Figure 2.10.

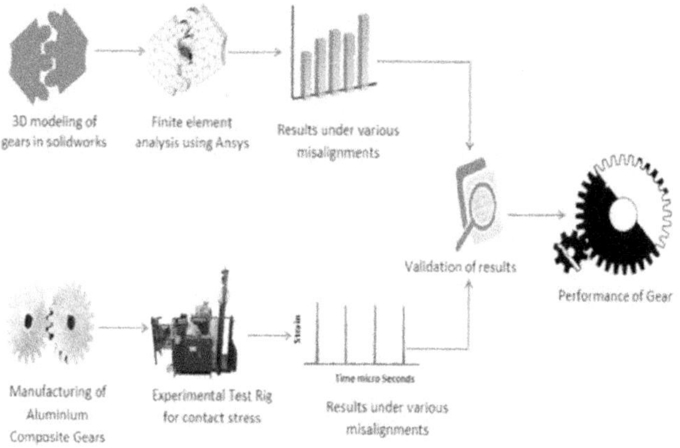

Figure 2.10. Block diagram of Phase-II work.

CHAPTER-3

3. ARCHITECTURE FOR IDENTIFICATION OF THE BEST MIXTURE

This objective deals with the mechanical and tribological behavior of different weight percentages of reinforcements (coated silicon carbide, boron carbide, and graphene oxide) dispersed with aluminum matrix (Al 6061) composites. The silicon carbide (SiC) was coated by titanium oxide (TiO_2) using the Electro Phoretic Deposition (EPD) method. The selected mix proportions fabrication process was performed using the stir casting method. The tribological and mechanical behavior of hybrid aluminum matrix-composite specimens and Scanning Electron Microscopy (SEM) identification revealed that ASB-1 (86 wt.% Al 6061 + 2 wt.% GO + 10 wt.% SiC + 2 wt.% B_4C) uncovered higher wear resistance and strengths due to reinforcement's uniform distribution throughout the aluminum matrix and coating of SiC. Moreover, a comparison was performed to determine the effectiveness of hybrid composite specimens. It is concluded that the presented SRNMS framework identified the mix proportions effectively, and the coating of SiC improves the wear resistance.

3.1. Materials and coating

A new generation of aluminum gears offers alternatives to more conventional metal gears. The Al 6061 is the higher strength-based aluminum alloy; hence, it was selected and used as a matrix material [89]. However, aluminum alloys have lower hardness and sometimes poor wear resistance. So, reinforcing the aluminum alloy with other materials can improve the mechanical behaviors of the material. Hybrid reinforcements such as SiC, B_4C, and GO have been selected to improve the mechanical properties.

3.1.1. Application of aluminum hybrid materials

Aluminum hybrid materials find applications across various industries due to their advantageous properties. In aerospace, they are utilized in aircraft fuselages, wings, and engine parts for their low density and high strength-to-weight ratio, reducing overall aircraft weight while maintaining structural integrity. The automotive industry benefits from aluminum's weight reduction capabilities, enhancing fuel efficiency and performance in engine components, body panels, wheels, and suspension parts. In construction, aluminum's corrosion resistance and lightweight nature make it ideal for windows, doors, roofing, and structural components. It is also a cost-effective choice for electrical conductors in power transmission lines and cables due to its balance of conductivity and weight. In packaging, aluminum's ability to resist corrosion and maintain the freshness of food and beverages makes it popular for cans, foil, and containers. The marine industry uses aluminum alloys for boat hulls, masts, and structural components due to their resistance to saltwater corrosion. Aluminum's excellent thermal conductivity is valuable in heat exchangers for automotive radiators and industrial equipment, enhancing energy efficiency. Sporting goods manufacturers use aluminum for lightweight yet strong products like baseball bats, bicycle frames, and tennis rackets. In aerospace, aluminum-lithium alloys reduce weight and increase fuel efficiency in aircraft and spacecraft. Electronics utilize aluminum for heat sinks and electromagnetic shielding, while medical devices benefit from its biocompatibility and lightweight properties. Architectural cladding uses aluminum composite panels for durability, aesthetics, and ease of maintenance. In transportation, aluminum improves fuel efficiency in trains, buses, and bicycles. It is valued for its strength-to-weight ratio in armored vehicles and defense equipment in military and defense. Renewable energy systems, such as wind turbines and solar panels, use aluminum for its lightweight and corrosion-resistant properties, enhancing system longevity and efficiency. Lastly, consumer goods like cookware, kitchen appliances, and furniture benefit from aluminum's attractive appearance, corrosion resistance, and ease of cleaning.

Due to the higher thermal conductivity, wear resistance, and smaller coefficient of thermal expansion, these materials have been selected. In addition, Al 6061 alloy is easily formed in nature, corrosion-resistant, and has moderate electricity. The properties of the selected reinforcement materials are shown in Table 3.1. Here, the selected alloy is thoroughly tested by the hardening process. Moreover, TiO_2 is chosen for the coating process. The properties of aluminum are shown in Table 3.2 [114].

Table 3.1. Reinforcement Materials properties

Product	Graphene Oxide (GO) nano powder	Silicon Carbide (SiC) nano powder	Boron Carbide (B_4C) nano powder
Purity (%)	99.99	99.90	99.90
Ammonium persulfate APS (nm)	10000	80-100	90
Molecular weight (g/mol)	12.01	40.11	55.26
Density (g/cm^3)	1.9-2.2	3.21	2.25
Melting point (^{0}C)	3452-3697	2820	2763
Solubility	Slightly soluble in water	Insoluble in water, acid, and alcohol	Insoluble in water

Table 3.2. The properties of Al 6061

Properties	Shear strength	Hardness	Fatigue strength	Elongation	Tensile strength	Elastic modulus	Poisons ratio
Values	207 MPa	95 HV	97 MPa	15%	310 MPa	70 GPa	0.33

3.1.2. Silicon carbide coating

In this research, TiO_2 is selected as a coating material for the SiC because it increases the bond strength between the matrix and reinforcements. The TiO_2 coating on SiC substrates was manufactured using the electrophoretic deposition method [115]. The TiO_2 specific surface area is 50 m^2/g, and the particle size is 21 nm. TiO_2 coats the SiC with 20 μm thickness; the components' optimal concentration in the solution is 0.04 g iodine in the 50 ml acetylacetone and 0.5 g TiO_2. The solution stability is improved after stirring sonically for 15 min. Hereafter, the solution was stirred magnetically for 3 minutes for each EPD test. After EPD, the coated SiC was carefully removed from the solution and dried at room temperature.

3.1.3. Manufacturing information

The manufacturing information for the SiC, B_4C, and GO materials are given below;

SiC: SiC is a complex chemical compound consisting of carbon and silicon; it is also a slight acid in chemical nature. Moreover, it is manufactured by mixing silica components and carbon at 250 °C temperature in an electric resistance furnace. Various manufacturing processes, such as large furnace technology and the traditional Acheson process, are used to manufacture SiC.

B_4C: Here, a heat resistance furnace is used in manufacturing, and petroleum coke and boric oxide are the raw materials for the B_4C. In this manufacturing process, a huge supply current passes through the graphite rod at the furnace's midpoint. The cylindrical furnace is coated with a mixture of boron oxide and coke. Based on the supply current, heat is created on the surface portion of the electrode. Then, the boron oxide is retorted with the coke to produce boron carbide material. SiC and B_4C are purchased in Morgan Advanced Materials, with the purity level for SiC being 99% or higher, while for B_4C, it is 95% or higher. B_4C and SiC sizes are 0.5 to 50 microns. B_4C and SiC particles can have different shapes, such as irregular or spherical.

GO: GO contains oxygen, hydrogen, and carbon, which are molecular suspensions of GO. Here, graphite is produced through various chemical methods. The manufacturing process of GO initially takes the artificially created graphite material with strong oxidizers such as sulphuric acids ($H_2 SO_4$). GO was purchased from ACS Material LLC. The purity level for GO can range from 90% to 99%. The size of GO is a few meters to several micrometers. GO particles can have different shapes, such as platelets or sheets. The behavior of GO particles after forming a hybrid composite, especially after the casting process, can vary depending on several factors, including the composite preparation method, processing conditions, and the interactions between GO and the matrix material (in this case, aluminum alloy). Here's an explanation of what might happen to GO particles in a hybrid composite;

Dispersion and distribution: In an ideal scenario, the GO particles should be well-dispersed and uniformly distributed within the aluminum matrix during the fabrication of the hybrid composite. Achieving good dispersion is crucial for maximizing the potential benefits of GO, such as enhanced mechanical and electrical properties. Techniques like sonication or the use of surfactants are often employed to promote uniform dispersion.

Size and shape stability: Generally, the size and shape of GO particles should remain relatively stable during the composite formation process. GO is typically in the form of sheets or flakes, and maintaining its two-dimensional structure is desirable. The processing conditions, including temperature and pressure, should be controlled to avoid significant changes in GO's size and shape.

Interfacial bonding: Depending on the surface functionalization of GO and the compatibility with the aluminum matrix, some level of chemical bonding or interactions may occur at the interface between GO and the matrix. This can lead to improved mechanical properties in the composite.

Possible reduction: In some cases, during the composite formation process, a reduction in GO may occur, leading to the formation of reduced graphene oxide. This reduction can be intentional or unintentional and may impact the electrical and thermal properties of the composite.

Post-casting analysis: Techniques like SEM and TEM were often employed to assess the behavior of GO particles after casting. These techniques can provide insights into the dispersion, distribution, size, and shape of GO within the composite. Changes in these characteristics can be observed through careful microstructural analysis.

3.2. Proposed SRNMS for mix selection

The Strawberry Recurrent Neural-based Mix Selection (SRNMS) is an advanced neural network framework specifically designed to address the challenges of composite mix selection. It leverages the capabilities of Recurrent Neural Network to analyze sequential dependencies and interactions between various components in a composite mixture. By integrating material properties, process parameters, and target performance metrics, SRNMS optimizes the selection and proportioning of components to achieve desired composite characteristics such as strength, hardness and wear resistance. The framework consists of multiple layers: an input layer that captures material data and constraints, a Recurrent Neural Network core that models interactions and dependencies over sequential additions, and an optimization layer that evaluates trade-offs between conflicting objectives like cost and performance. The decision layer provides actionable recommendations, offering the best combinations for specific applications while dynamically adapting to changing constraints. This system is particularly valuable in material science, manufacturing, and environmental engineering, enabling the creation of high-performance, cost-effective, and sustainable composites. By reducing trial-and-error in traditional design methods and ensuring data-driven decisions, SRNMS represents a transformative approach to composite development.

The main motive of this present research work is to improve aluminum gears' tribological and mechanical properties and to identify high-strength hybrid nanocomposite specimens [116]. To minimize the cost and time, a novel SRNMS technique is utilized to discover the high-strength mix proportion. Here, reinforcement mixing proportions are the input material, and these materials are initiated to the neural network's input layer. Based on these mixing proportions, the correspondent hybrid composites are created. Moreover, an optimization fitness function is used to prepare the high-strength hybrid nanocomposite specimens to optimize the tribological behaviors. The layers of SRNMS are shown in Figure 3.1.

So, here, a novel Strawberry optimization is chosen to determine the effective weight percentage of reinforcements. Different filler percentages with equal intervals are used as an additive in the aluminum material and implemented in MATLAB to find the adequate mix proportion. Finally, the testing outcomes were attained and validated with other mixes to justify the optimal state of the selected best mix. Using this approach, the system can optimize the proportions of the composite's reinforcing materials, coated silicon carbide, boron carbide, and graphene oxide, to achieve the desired mechanical and tribological properties. The Recurrent Neural-based Network (RNN) can predict the properties of the composite material based on the input variables, such as the proportions of the reinforcing materials. At the same time, the Strawberry optimization algorithm can find the optimal combination of the input variables that maximize the desired properties. This approach can lead to more efficient and cost-effective manufacturing of composite materials with optimized properties. It can reduce the need for trial-and-error experimentation and minimize the materials used. It can also help improve composite materials' performance and reliability in various applications, such as the aerospace, automotive, and construction industries.

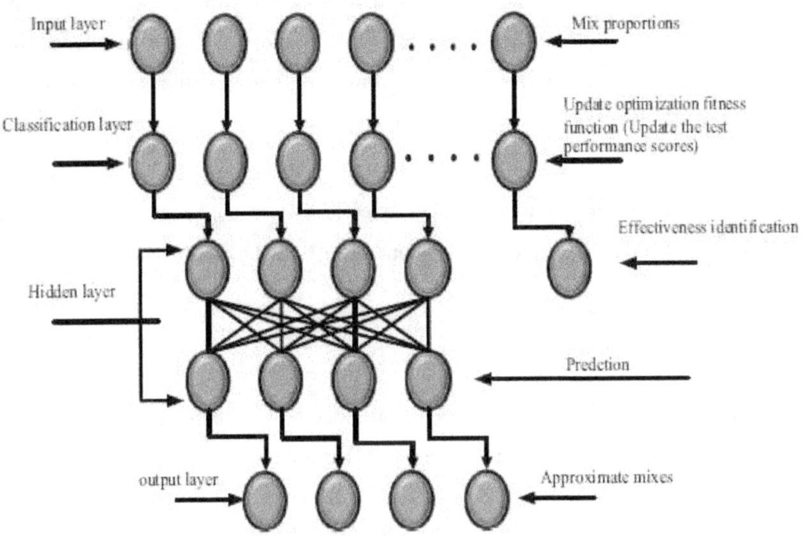

Figure 3.1. Layers of SRNMS.

3.2.1. Process of SRNMS

In this work, the SRNMS are incorporated for mix proportion selection by optimization, and it is developed as a new framework. The study used the Recurrent Neural-based Network to predict the Hybrid Nanocomposites' properties based on the selected mix proportions. The properties of the composite were predicted using the RNN trained on the data obtained from a set of experiments conducted on the composites. The aim was to create a model that could accurately predict the properties of the composites for any given mix proportions without the need for actual experiments. Various mixtures are required for trial-

and-error analysis in specimen preparation, increasing materials usage, preparation time, and work cost. To avoid these disadvantages, a novel SRNMS was chosen to select the appropriate mix proportion based on the fitness function of berries. The chosen mix of proportioning data containing 26 samples was initially trained to the MATLAB system by Equation 3.1.

$$Q(m) = M_p\{1, 2, 3, \dots, n\} \qquad \text{.................. } \textit{Equation (3.1)}$$

Here, Q(m) represents the training parameter of mix proportions, and Mp represents the input mix proportions. Moreover, the given mix proportions are given to the hidden layer after training. In the hidden layer, the strawberry fitness is utilized to identify the adequate mix proportions and predict the appropriate ones.

There are four layers in the presented SRNMS: the input layer, two hidden layers, and the output layer. In the hidden layer, there are two functions, i.e., mix effectiveness identification and prediction. The neural layer training was done using Equation 3.2.

$$M_t(b, p, fs) = -\sum_{l,r} b_l . p_r . W_{lr} - \sum_{l} u_l . v_l - \sum_{r} p_r h_r + \sum_{l,r} (b + p) fs \qquad \text{............ } \textit{Equation (3.2)}$$

Where l and r represent the visible and hidden layer, the invisible and visible units are defined as b, p, and W, which is the weight function. The remote and visible layer variables are h and v, respectively. In addition, neural layer training is denoted as M_t, and f_s is strawberry fitness. Moreover, the effective identification of mixed proportions was performed by Equation 3.3.

$$E_{fi} = M_p + e_f \times rand(d, s) \Rightarrow E_{fim} \qquad \text{............... } \textit{Equation (3.3)}$$

E_{fi} is the effectiveness identification module; e_f represents the searching layer; (d, s) is the minimum and maximum strengths allocated for identification; here, d = 124 MPa, and s = 290 MPa. Specimens with tensile strength less than or equal to 124 MPa are considered as normal, specimens with tensile strength lying in between 124 MPa to 290 MPa are considered as neutral and specimens with tensile strength greater than 290 MPa are considered as good. Consequently, the prediction of correct mix proportions was performed by Equation 3.4.

$$P_{m_a}^{*} = \begin{cases} \frac{1}{E_{fi}+f(E_{fi}.E_{fim})} f(E_{fim}(i)) & = \leq \ 124MPa, normal \\ E_{fi} + f(E_{fi}.E_{fim}) f(E_{fim}(i)) & = \leq \ 290MPa, \ Neutral \\ E_{fi} + f(E_{fi}.E_{fim}) f(E_{fim}(i)) & = > 290MPa, \ Good \end{cases} \quad \text{.................. Equation (3.4)}$$

Here, E_{fim} represents the effectiveness of the identified mix, and $P_m{}^*{}_a$ predicts appropriate mixes. The correct mix proportions with higher strengths are output in the output layer.

The output attained from the SRNMS framework is utilized for specimen preparation. In addition, the pseudocode of the presented SRNMS model is shown in Algorithm 1. Moreover, the overall workflow procedure is indicated in the flowchart (as shown in Figure 3.2).

The composite gears' effectiveness is measured in terms of flexural strength, impact strength, tensile strength, friction coefficient, wear rate, abrasion, and worn surfaces. The evaluation results selected the compelling mix for the gear manufacturing process.

Algorithm 1: SRNMS-based mix proportion selection

Start

{

Initialize: *a, s,* *// materials initialization*

b, g

//Here, a is the Al 6061, s is the silicon carbide, b is the boron carbide, and g is the graphene oxide

Design: SRNMS

$$Q(m) \Rightarrow M_p$$

// mix proportions training

$M_t \rightarrow M_p$

// neural layer training

Effectiveness identification ()

{

Allocation of strengths

//Here, the range of strength allocated is 124MPa and 290MPa

$E_{fi} \rightarrow E_{fim}$ ***// strawberry fitness updation***

//Here E_{fi} *is the effective mix identification parameter*

}

Prediction of appropriate mixes ()

{

$P^{*}_{ma} \rightarrow O_{cm}$

//Here, O_{cm} *is the predicted appropriate mixes*

Conditions for predictions

$E_{fim}(i) = \leq 124MPa$

// this condition is satisfied, and then it is not selected for specimen preparation

$E_{fim}(i) = \leq 290MPa$

//It is suitable for specimen preparation

$E_{fim}(i) = > 290MPa$

//It is highly suitable for specimen preparation

}

Appropriate mix proportions ()

{

$O_{cm} \Rightarrow A_{M_p}$

//Here, A_{M_P} *is the appropriate mix proportions for specimens' preparation*

}

Stop

}

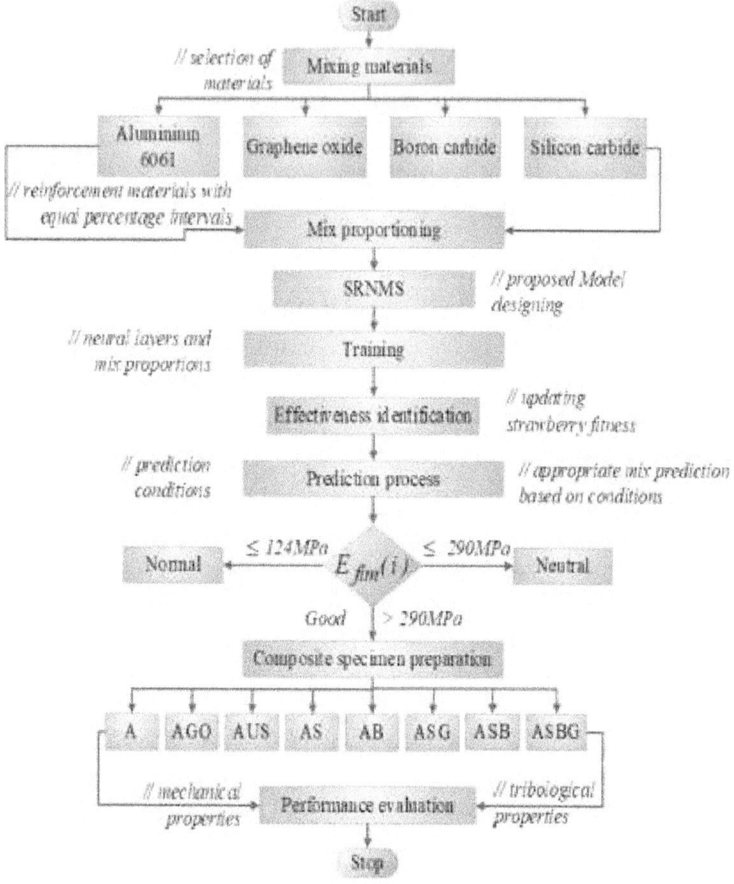

Figure 3.2. Overall workflow process.

3.3. Methodology

3.3.1. Fabrication of composites

The fabrication of composites using the stir-casting technique was a meticulous process designed to ensure the uniform distribution of reinforcements within the matrix. This method was chosen for its efficiency in achieving a homogeneous composite structure. The process began with selecting a furnace capable of reaching a maximum temperature of 1000°C, suitable for melting and processing the required materials. A graphite crucible with a 2 kg capacity was selected to match the furnace's specifications to ensure optimal thermal conductivity and stability. The primary matrix material, Al 6061-T6 aluminum alloy, sourced from Bharat Aerospace Metals Pvt. Ltd, India, was placed in the graphite crucible. The T6 process involves solution annealing, quenching, and artificial aging to enhance aluminum alloys' properties. Solution annealing dissolves alloying elements at high temperatures, followed by rapid quenching to trap them in a supersaturated state. Artificial aging at controlled temperatures precipitates strengthening phases, improving hardness, tensile strength, and overall mechanical performance. This aluminum alloy was chosen for its excellent mechanical properties and broad applicability in various industries. The melting process was conducted at 850°C, higher than aluminum's melting point. This elevated temperature was crucial for removing impurities and ensuring a clean, high-quality melt. Simultaneously, the reinforcement materials, including β-SiCp from US Research Nanomaterials, Inc., USA, Graphite from Mknano, M K Impex Corp., Canada, and Boron Carbide from NanoShel, Corp., UK, were preheated to 850°C. Preheating the reinforcements to the same temperature as the molten aluminum improved their wettability and integration into the matrix. The photographs of Al 6061 and powders are shown in Figure 3.3, and their details are shown in Table 3.3.

Metallic molds were prepared by preheating them to 500°C to prevent thermal shock and ensure smooth casting. Once the aluminum reached 850°C and was impurity-free, it was extracted from the graphite crucible and transferred to an electrical furnace for mixing. As illustrated in Figure 3.4, the stir-casting setup employed precise stirring parameters to achieve a uniform composite. The stirring was performed at 500 rpm for 6 minutes using a mechanical stirrer, ensuring thoroughly mixing of the aluminum and the reinforcement particles. This step was critical for avoiding particle agglomeration and ensuring an even distribution within the molten matrix.

After stirring, the molten composite mixture was carefully poured into a preheated cast-iron mold with 150 mm x 90 mm x 10 mm dimensions. The mold was designed to shape the composite while allowing it to cool gradually at room temperature, preventing any thermal stresses that could lead to defects. Once solidified, the composite was removed from the mold and machined to the required ASTM standard dimensions for subsequent testing. The fabricated samples underwent mechanical tests to evaluate properties such as tensile strength, hardness, and wear resistance. The photographs of fabricated composites are shown in Figure 3.5.

Figure 3.3. The photographic images of (a) Al 6061 aluminum alloy, (b) β-SiCp, (c) Graphite and (d) Boron Carbide

Table 3.3. Details of materials and powders

	Al 6061-T6	β-SiCp (silicon carbide)	Graphite	Boron Carbide
Color	Silver white	Grayish white	Black	Black
Density (g/c.c)	2.7	3.2	2.2	2.5
Morphology	FCC	Cubic	Hexagonal	Complex

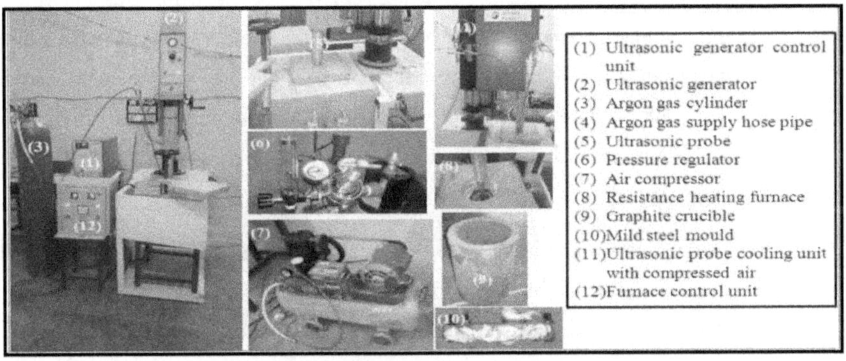

Figure 3.4. Stir casting technique.

Figure 3.5. Output from stir casting

3.3.2. Microstructural investigation

The microstructures of the specimens, captured using a scanning electron microscope (SEM-ZEISS model), reveal the materials' diffusivity. Additionally, an energy-dispersive X-ray analysis (EDAX) is extensively used to assess the chemical composition of the composite material by analyzing SEM images, providing detailed elemental information.

3.3.3. Tensile testing

The tensile strength of the composites was determined using the Universal Testing Machine (UTM) following ASTM E8 standards. Specimens with dimensions of 130 mm in length, 25 mm in width, and 5 mm in thickness were subjected to uniaxial tension until failure. This test measured the material's ultimate tensile strength, yield strength, and elongation, essential for understanding its behavior under tensile loads.

3.3.4. Flexural testing

Flexural strength was assessed following ASTM D790 standards. Specimens measuring 115 mm in length, 20 mm in width, and 5 mm in thickness were tested using a three-point bending setup. The UTM applied force at a constant rate until the specimen fractured. The test determined the material's ability to resist deformation under load, providing insights into its bending strength and stiffness.

3.3.5. Hardness test

The hardness test was carried out to analyze the effect of filler material on the composites using a well-defined methodology to ensure consistent and accurate results. Composite samples were prepared with smooth surfaces and calibrated using a Brinell hardness testing machine (Model: ECNOMET) equipped with a 5 mm diameter steel ball indenter applied a 3000 N load for 12 seconds. The indentation left on the composite surface was measured, and this process was repeated multiple times to ensure reliability, with average measurements taken to account for variability. The hardness test was conducted as per the ASTM E10 standards.

3.3.6. Impact testing

The impact strength of the composite specimens was evaluated using the Charpy impact testing machine, adhering to ASTM E23 standards. Specimens were prepared with dimensions of 55 mm in length, 20 mm in width, and 10 mm in thickness. The Charpy impact test involved striking the specimen with a pendulum hammer to measure the energy absorbed during fracture, indicating the material's toughness and resistance to impact forces.

3.3.7. Wear test

According to ASTM standards, the specimens for wear analysis are cylindrical dies measuring 175 mm in length and 20 mm in diameter. The wear rate, which indicates the mass loss ratio, material density, load, and sliding distance, is evaluated using Equation 3.5:

$$\text{Wear rate} = \frac{\Delta m}{sd\rho} \times 10^3 \quad \text{……………………. Equation (3.5)}$$

Where Δm is the mass loss, s represents the load, d is the sliding distance, and ρ is the density of the test specimen. The unit of wear rate is mm^3/Nm. The dry sliding wear properties of composites were analyzed using the pin-on-disc method as per ASTM G99. The disc material is EN-32 steel, with loads of 10 N and 20 N under dry sliding conditions. The sliding distance and track diameter are both 100 nm. Additionally, the disc and pin wear samples were polished with emery papers.

3.4. Results and discussion

3.4.1. Optimization

This investigation evaluates the morphological, mechanical behavior, and tribological of the Al 6061 + coated/uncoated silicon carbide + graphene oxide + boron carbide hybrid composites. A novel SRNMS is developed for the mix proportioning section. The stir casting technique was utilized for fabrication, and ASTM standard specifications were followed for specimen preparation and testing. It is well-known that nature-inspired optimization can tackle numerous complicated engineering issues compared to traditional optimization, such as gradient-dependent optimization. Several heuristic optimizations are available for the mathematical solution; still, the strawberry optimization has better fitness than the other optimization running roots. Here, it helps to analyze all mix samples at every iteration, and the best selection mix was found based on different tests such as flexural strength, impact strength, tensile strength, hardness, wear rate, and friction coefficients.

Several mix ratios are taken to estimate the performance of the developed SRNMS framework. Initially, Al 6061, coated silicon carbide, boron carbide, and graphene oxide are selected with different mix proportions. The different ratio of mix proportions implemented in MATLAB is shown in Table 3.4. Moreover, the chosen mix proportions are trained to the MATLAB R2020a and run on the Windows 10 platform, and the presented method is designed in the MATLAB environment. Hereafter, the neural layers of the designed SRNMS framework are trained.

In this process, 26 mix proportions were selected and trained for the system, in which the A-1 mix was taken as a base mix proportion for further processing. The fitness of strawberries is updated in the effectiveness identification layer, which determines each mix proportion's effectiveness. In the prediction layer, the conditions were applied to determine the appropriate mix proportions. Moreover, the SRNMS-identified mix proportions are represented in Table 3.5. The finest optimal mix specimen is ASB-1. Based on the best strength value, the best mix selected was ABS-1.

The mixed proportions selected by the presented framework are used for preparation purposes. In preparation, aluminum is melted in the crucible, and the stir-casting method prepares the specimens. As per ASTM standards, the specimen preparation process and dimensions of samples are followed. The tensile, flexural, impact strength and hardness tests show that the optimized mix selection process determines the mix proportion effectively. The parameters utilized in the wear and coefficient of friction and the abrasive test are given in Table 3.6.

The composite specimens are transferred to cast-iron mold for shaping. Hereafter, the tribological and mechanical properties are evaluated. The test results selected the best compositions of Al 6061 with reinforcements for the gear manufacturing process. Moreover, to determine the presented framework SRNMS, the input mix proportion ASG-1 (91 wt.% Al 6061 + 2 wt.% GO +5 wt.% SiC +2 wt.% B_4C) is taken for tensile strength evaluation.

Table 3.4. Mix proportions chosen for implementation

S. No	Specimen code	Al 6061	GO	SiC	B_4C	Test specimen material
1	A-1	100	0	0	0	Pure Al 6061
2	AGO-1	95	5	0	0	Al 6061 with GO reinforcement
3	AGO-2	90	10	0	0	
4	AGO-3	85	15	0	0	
5	AGO-4	80	20	0	0	
6	AUS-1	95	0	5	0	Al 6061 with uncoated SiC reinforcement
7	AUS-2	90	0	10	0	
8	AUS-3	85	0	15	0	
9	AUS-4	80	0	20	0	
10	AS-1	95	0	5	0	Al 6061 with coated SiC reinforcement
11	AS-2	90	0	10	0	
12	AS-3	85	0	15	0	
13	AS-4	80	0	20	0	
14	AB-1	95	0	0	5	Al 6061 with B_4C reinforcement
15	AB-2	90	0	0	10	
16	AB-3	85	0	0	15	
17	AB-4	80	0	0	20	
18	ASG-1	91	2	5	2	Al 6061 + 5% coated SiC + GO and B_4C
19	ASG-2	89	3	5	3	
20	ASG-3	87	4	5	4	
21	ASB-1	86	2	10	2	Al 6061 + 10% coated SiC + GO and B_4C
22	ASB-2	84	3	10	3	
23	ASB-3	82	4	10	4	
24	ASBG-1	81	2	15	2	

25	ASBG-2	79	3	15	3	Al 6061 + 15% coated SiC + B_4C and GO
26	ASBG-3	77	4	15	4	

Table 3.5. Prepared composite specimens obtained strengths

Specimen code	Flexural strength (MPa)	Impact strength (J)	Tensile strength (MPa)	Hardness (HB)	Wear rate (mm^3/Nm) $\times 10^{-3}$	COF	AFC
A-1	160.64	22.75	292.89	265.47	8	0.9	0.83
AGO-1	159.04	20.02	270.7	238.73	8.2	0.93	0.845
AGO-2	158.4	21.67	260.48	240.99	8.5	0.923	0.831
AGO-3	173.42	28.44	311.7	310.12	7.7	0.86	0.79
AGO-4	159.95	19.09	290.5	231.22	8.42	0.941	0.88
AUS-1	157.3	17.01	284.4	216.18	8.68	0.982	0.87
AUS-2	179.5	32.54	315.54	296.25	7.5	0.80	0.70
AUS-3	156.8	15.09	288.75	243.98	8.29	0.919	0.878
AUS-4	158.9	15.69	250.5	261.22	8.34	0.972	0.91
AS-1	155.8	17.03	265.2	234.76	8.56	0.989	0.93
AS-2	154.3	16.78	278.34	244.54	8.81	0.961	0.971
AS-3	156.23	18.3	269.79	243.2	9.1	0.934	0.846
AS-4	164.72	34.92	316.72	270.37	7.9	0.79	0.75
AB-1	161.94	25.79	309.84	289.4	6.5	0.77	0.61
AB-2	154.22	16.02	288.59	250.23	8.99	0.958	0.875

AB-3	156.9	18.55	269.89	241.12	8.72	0.965	0.889
AB-4	157.58	16.45	285.95	253.24	8.36	0.993	0.839
ASG-1	158.67	17.5	280.89	129.34	8.49	0.915	0.863
ASG-2	155.98	20.89	270.99	212.11	8.56	0.951	0.891
ASG-3	183.25	26.5	309.89	300.25	6.7	0.72	0.55
ASB-1	200.75	36	325.6	350.45	5.54	0.46	0.49
ASB-2	156.37	20.44	262.99	232.87	8.92	0.976	0.945
ASB-3	155.67	17.24	278.43	223.34	8.61	0.989	0.923
ASBG-1	154.56	16.12	256.78	210.89	8.71	0.948	0.94
ASBG-2	189.52	29	307.41	320.45	7.3	0.7	0.60
ASBG-3	152.22	21.22	245.55	220.12	9.21	0.99	0.95

Table 3.6. Wear and coefficient of friction and abrasive test parameters

Parameters	**Load applied (N)**	**Sliding Speed (m/s)**	**Sliding distance (m)**
Range of values	10, 20, 30 and 40	1.04, 1.57, 2.09 and 2.61	500, 1000, 1500 and 2000

3.4.2. Microstructural behavior of the materials

The specimens' microstructure images show the materials' diffusivity using an SEM. An EDAX is a broadly utilized method to assess the chemical elements of composite material using SEM images. The testing output in both SEM and microscopy images is shown in Figure 3.6. The microstructural analysis supported the claim that the ASB-1 (Al 6061 + 10% coated SiC + 2% GO and 2% B_4C) composite material can enhance mechanical and tribological properties and increase its lifetime. Microstructural analysis can provide insights into the composite material's microstructure, which can then be correlated to its microscopic, mechanical, and tribological properties. Microstructural analysis showed that the SiC, GO, and B_4C reinforcements were uniformly distributed throughout the aluminum matrix, with good bonding between the matrix and the reinforcement particles. This uniform distribution and good bonding can improve the composite material's mechanical and tribological properties. It may have also revealed that the composite material exhibited less wear and damage than the matrix material. This observation can support the claim that the composite material can increase the material's lifetime. Hence, it was proved that the selected sample could enhance the mechanical and tribological properties and increase the material's lifetime.

Moreover, EDS mapping analysis can analyze the elemental composition of aluminum-based materials, such as the Al 6061-SiC-B_4C-GO hybrid nanocomposites discussed in the present study. By using EDS mapping, it is possible to determine the relative concentrations of aluminum, silicon, carbon, boron, and any other elements in the composite. EDS mapping analysis is also taken for aluminum here, and red dots indicate improved visibility criteria that are manually enlarged with suitable coating composites such as 86 wt.% Al 6061 + 2 wt.% GO +10 wt.% SiC +2 wt.% B_4C. These areas of interest correspond to regions where the coating composites are. The coating composites may have improved the mechanical, thermal, or other properties of the Al 6061-SiC-B_4C-GO hybrid nanocomposites. By using EDS mapping, it is possible to confirm the presence and relative concentrations of the various elements in the coating composites and to verify that they have been successfully applied to the sample. Consequently, the EDS analysis is shown in Figure 3.7.

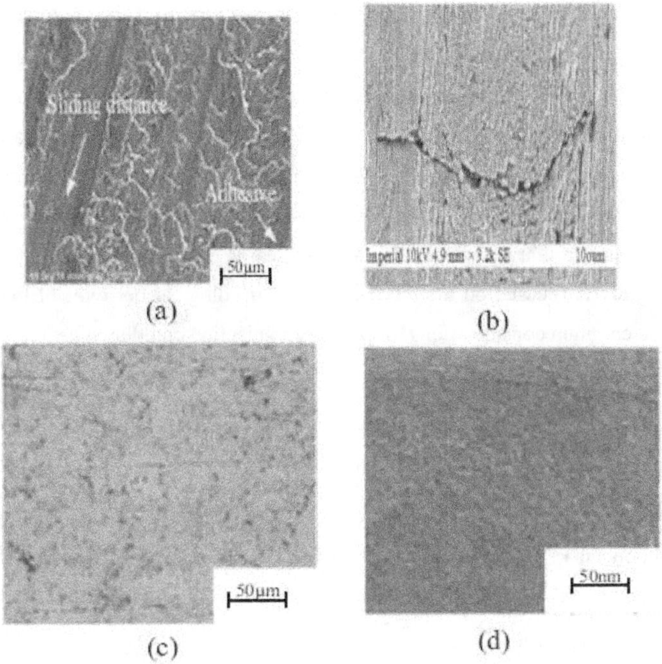

Figure 3.6. Testing output in SEM and microscopy images: a) SEM results, b) RCF test output, c) microstructure image using an optical microscope, d) surface roughness using Mitutoyo company surtronic 3+ measuring tool.

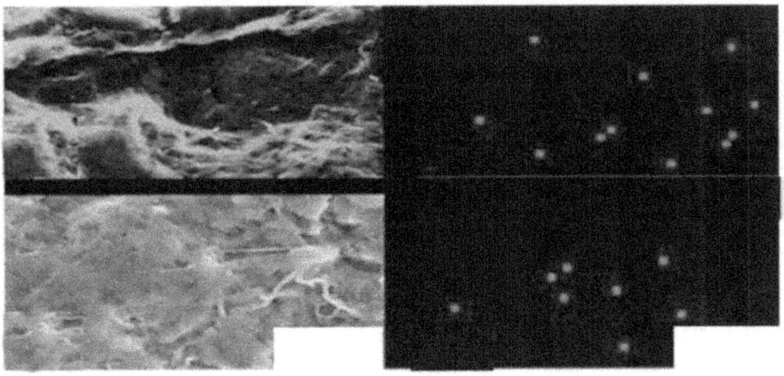

Figure 3.7. EDS results.

3.4.3. Wear rate with varying load, sliding distance, and sliding Speed

The specific parameters for tribological testing, including the load, sliding speed, and sliding distance, are typically selected based on several factors, including the intended application, the tested materials, and the conditions that best simulate the real-world tribomechanical system under observation. These parameters are chosen to replicate the operating conditions as closely as possible. The real tribomechanical system under observation might vary depending on the research or application, but in this context, where the study is focused on the mechanical and tribological behavior of hybrid composites, some common considerations for parameter selection could include:

Load: The load applied during tribological testing is chosen to mimic the mechanical stresses experienced by the material in its intended application. It should consider factors like the weight or force the material will bear in real-world use.

Sliding speed. Sliding speed is selected to replicate the relative motion between components in the actual system. It is chosen based on the expected operating speed of the components and can range from very slow to high-speed conditions.

Sliding distance. The anticipated travel or movement of the components in the real system determines the sliding distance or stroke length. It is essential to consider how far the components will slide or interact during their intended use. In this case, the real tribomechanical system under observation appears to involve components or materials that experience sliding or relative motion against each other, and the study aims to understand how the hybrid composites behave under these conditions. The specific parameters for load, sliding speed, and sliding distance would have been chosen to replicate the operational conditions expected in the real system, allowing researchers to accurately assess the composite materials' tribological performance.

The mean weight loss is utilized to calculate the specific wear rate. The abrasive wear tests changed the average load, sliding distance, and speed. In addition, to identify the wear mechanism and wear type, SEM images were taken to identify the worn surface. SEM analysis provides high-resolution images of the sample's surface, which can be used to observe the morphology and distribution of various components. It is also possible that the uniform distribution of reinforcements was not visually apparent in a single image but inferred from statistical analysis or other methods. The rate of wear related to the different sliding distances ranges from 500 to 2000 m. Slip systems are defined as the deformation of the material while applying the loads. Using the sliding distance, the deformation of the materials is evaluated. Figure 3.8(a) denotes the wear rate related to the different load conditions from 10 to 40 N for various composite specimens. As the graph observes, the wear rate gradually increases while adding the load levels ranging from 10 N to 40 N. From the obtained results, the ASB-1 specimen attained a lower wear rate at a maximum load application of about 40 N compared with other samples. The wear rate is lower in the ASB-1 specimen compared to other models observed in Figure 3.8(b).

The wear rate of the composite materials increases with the increase in sliding distance. The rate of wear reduction is a more advantageous parameter by creating a protection phase. The reinforced composite specimen provides a better barrier by restricting delamination, consequently decreasing the loss of materials. In addition, Figure 3.8(c) indicates the changes in wear rate related to the different sliding speeds ranging from 1.04 to 2.61 m/s. The standard deviation of the average wear rate values with load, sliding distance, and sliding velocity are shown in Figure 3.9.

$$R_{S/N} = -10 \log \frac{1}{n} \sum_{j=1}^{n} H_j^2 \qquad \text{... ... ... Equation (3.6)}$$

The S/N ratio is $R_{S/N}$ given by the above formula (Equation 3.6) using logarithm terms, taking the iteration of each value n with the output H^2_j of the jth experiment for each trail. The variables are considered for statistical analysis when predicting values or making decisions. After getting the parameters, techniques are used to get less wear rate.

$$S_{wr} = M_{wr} + (F_2 - M_{wr}) + (R_1 - M_{wr}) + (K_1 - M_{wr}) \qquad \text{... ... ... Equation (3.7)}$$

The above Equation 3.7 gives the specific wear rate S_{wr} by combining the mean value of wear rate in total M_{wr} and subtracting M_{wr} from F_2, R_1, and K_1 main factors at selected levels.

In conformation analysis, a particular confidence interval estimates a set of values to check its quality. This interval is chosen chiefly with 95% and gives an exact value within the range. By comparing intervals to quality standards, the characteristics are observed.

$$C_I = \sqrt{E_{\alpha;1,v_2} \bullet V_e \bullet \left(\frac{1}{\eta_{eff}} + \frac{1}{r}\right)} \quad \text{.................................. } \textit{Equation (3.8)}$$

In the above Equation 3.8, The $E_{\alpha:1;v2}$ distribution value obtained from the statistical table depends on 1; V_2 and confidence chosen atα. V_2 is the total degrees of freedom associated with the variance of the error term in an analysis of variance or regression model. V_e is a measure of the variation or error within the data. r indicates how many times a particular measurement or observation was taken. η_{eff} is often used in the context of weighted averages (as represented in Equation 3.9).

$$\eta_{eff} = \frac{\text{number of trails}}{1 + \text{degree of chosen parameters}} \quad \text{........................... } \textit{Equation (3.9)}$$

Additionally, the wear rate increases while the sliding speed increases. It occurs because the temperature rises, which softens the contact surface and enhances the wear rate. The graph showed that the ASB-1 specimen attained a lower wear rate when compared to the other samples. Wear mechanisms refer to the various processes by which materials degrade and lose mass when they are in contact. Wear can occur in mechanical components, industrial machinery, automotive parts, and other contexts where materials interact. There are several primary wear mechanisms:

Abrasive wear. This occurs when hard particles or abrasive substances come into contact with a softer material, causing the removal of material from the softer surface. It's similar to sandpaper wearing away a surface.

Adhesive wear. Two surfaces come into contact in adhesive wear, and micro-welding can occur between them. When they separate, material from one surface can be transferred to the other, causing wear. It's often called galling.

Erosive wear. It happens when high-velocity particles or fluids impinge on a surface and remove material. Examples include erosion of aircraft surfaces due to rain and sand particles.

Corrosive wear. Corrosion wear is a combination of wear and chemical corrosion. It occurs when a material degrades due to chemical reactions with substances in its environment, leading to material removal.

Fatigue wear: Fatigue wear results from repeated cyclic loading and unloading of a material, leading to cracks and eventual material removal. This is common in applications subject to repeated stress, like rotating machinery. Understanding these wear mechanisms is essential in engineering and materials science to design components that can withstand wear and extend their lifespan. Strategies like using wear-resistant coatings, selecting appropriate materials, and implementing lubrication and maintenance practices are often employed to mitigate wear and its detrimental effects [117].

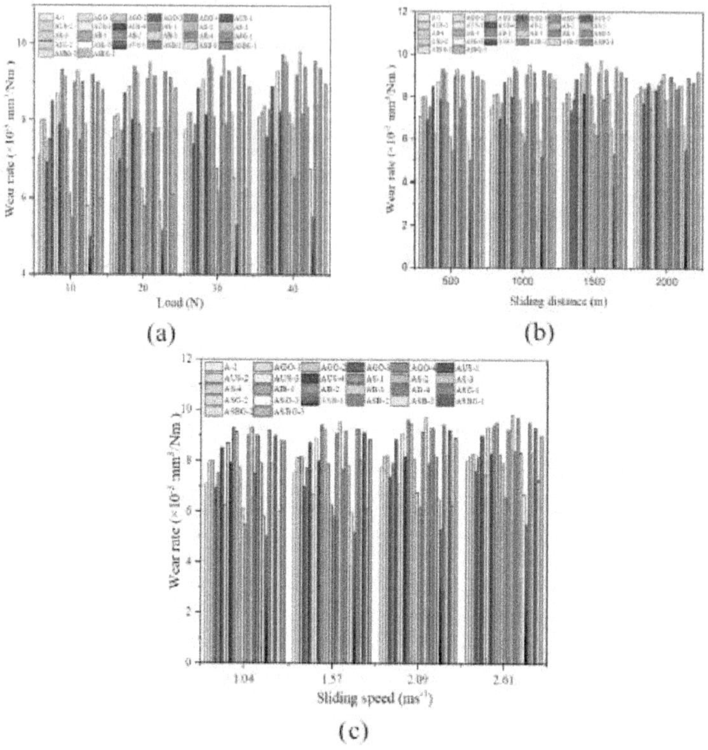

Figure 3.8. Obtained wear rate (a) with varying load, (b) with varying sliding distance, and (c) with varying sliding Speed.

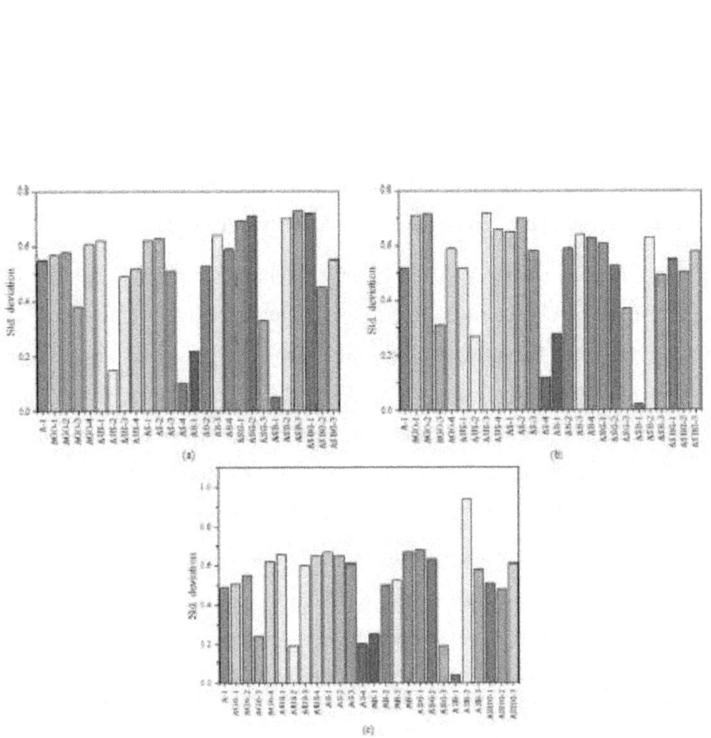

Figure 3.9. Average outcomes with error bar standard deviation for wear rate (a) load, (b) sliding distance, and (c) sliding speed.

3.4.4. The friction coefficient and abrasive friction coefficient (AFC)

The friction coefficient is a significant parameter for evaluating the gear's surface durability and scoring resistance. The friction coefficient is the calculation of friction between two contact surfaces. The friction force is applied and slides against the motion against both contact surfaces. Therefore, it is necessary to investigate its effect on the reinforcement materials and their friction-wear properties in the case of adhesive and abrasive wear. Abrasive, in which a harder element eliminates material from a softer one. The friction coefficient is calculated using Equation 3.10.

$$COF = \frac{F.F}{N.F} \qquad \text{.........................} \textit{Equation (3.10)}$$

Here, COF indicates the friction coefficient, F: F signifies the friction force in N, and N: F denotes the normal force in N. The wear rate with sliding wear and its ratio are mentioned in Table 3.7. The COF on differing sliding speeds is defined in Table 3.8. The ASB-1 specimen attains a lower friction coefficient when compared with other samples. It shows that the ASB-1 specimen achieves a lower friction coefficient.

In addition, the friction that causes rotor spinning is determined as abrasive. In a moving vehicle, the two brake pads were pressed together, and then sufficient friction was gained to stop the rotor. Because it was abrasive, a wide range of heat was generated, which could soften the rotor. It has tended to cause particles to break down from the brake pad. Finally, the performance of the rotor and brake pads failed, causing the gear failure mode. Figure 3.10(a) shows the failure mode of all mixes, and Figure 3.10(b) shows the time taken for the best specimen. The optimal mix ASB-1 has gained the lowest displacement at every stage with different loads. The high AFC can cause a high failure-mode rate.

Table 3.7. Wear rate with sliding wear and its ratio.

S. No	Weight (%)	Load (N)	Sliding Speed (m/s)	Sliding Wear rate x 10^{-5} mm^3 /(Nm)	S/N Ratio of Specific Wear rate (dB)
1	3	10	1.04	1.580	−3.9731
2	3	10	1.39	2.540	−8.0967
3	3	10	2.61	3.230	−10.1841
4	3	20	1.04	3.785	−11.5613
5	3	20	1.39	6.055	−15.6423
6	3	20	2.61	7.720	−17.7523
7	3	30	1.04	4.313	−12.6963
8	3	30	1.31	5.183	−14.2922
9	3	30	2.61	6.853	−16.7180
10	5	10	1.04	1.110	−0.9065
11	5	10	1.31	1.610	−4.1365
12	5	10	2.61	2.450	−7.7833
13	5	20	1.04	3.265	−10.2777
14	5	20	1.31	5.655	−15.0487

15	5	20	2.61	6.460	−16.2047
16	5	30	1.04	3.846	−11.7017
17	5	30	2.61	4.510	−13.0835
18	5	30	1.31	6.003	−15.5678

Table 3.8. COF based on sliding speed

Sliding speed	**1.04**		**1.57**		**2.09**		**2.61**	
	COF	**AFC**	**COF**	**AFC**	**COF**	**AFC**	**COF**	**AFC**
A-1	0.72	0.71	0.76	0.74	0.82	0.79	0.9	0.83
AGO-1	0.81	0.73	0.83	0.75	0.88	0.79	0.93	0.845
AGO-2	0.82	0.75	0.86	0.79	0.9	0.802	0.923	0.831
AGO-3	0.8	0.71	0.82	0.73	0.84	0.76	0.86	0.79
AGO-4	0.86	0.79	0.9	0.81	0.91	0.85	0.941	0.88
AUS-1	0.9	0.8	0.92	0.82	0.95	0.84	0.982	0.87
AUS-2	0.71	0.61	0.73	0.63	0.77	0.67	0.8	0.7
AUS-3	0.85	0.81	0.88	0.84	0.902	0.86	0.919	0.878
AUS-4	0.907	0.83	0.93	0.87	0.956	0.89	0.972	0.91
AS-1	0.891	0.84	0.93	0.87	0.961	0.911	0.989	0.93
AS-2	0.91	0.903	0.93	0.925	0.952	0.956	0.961	0.971
AS-3	0.88	0.72	0.89	0.76	0.91	0.81	0.934	0.846
AS-4	0.71	0.69	0.73	0.71	0.76	0.73	0.79	0.75
AB-1	0.71	0.51	0.73	0.53	0.75	0.58	0.77	0.61

AB-2	0.9	0.81	0.92	0.84	0.941	0.86	0.958	0.875
AB-3	0.91	0.804	0.93	0.83	0.947	0.851	0.965	0.889
AB-4	0.934	0.751	0.958	0.783	0.972	0.815	0.993	0.839
ASG-1	0.82	0.81	0.85	0.83	0.87	0.85	0.915	0.863
ASG-2	0.911	0.808	0.922	0.826	0.941	0.871	0.951	0.891
ASG-3	0.61	0.46	0.65	0.48	0.68	0.51	0.72	0.55
ASB-1	0.402	0.401	0.421	0.422	0.442	0.462	0.46	0.49
ASB-2	0.91	0.9	0.94	0.91	0.96	0.93	0.976	0.945
ASB-3	0.92	0.83	0.94	0.85	0.96	0.88	0.989	0.923
ASBG-1	0.924	0.903	0.934	0.913	0.952	0.934	0.98	0.94
ASBG-2	0.61	0.51	0.63	0.53	0.67	0.58	0.7	0.6
ASBG-3	0.91	0.89	0.94	0.91	0.97	0.93	0.99	0.95

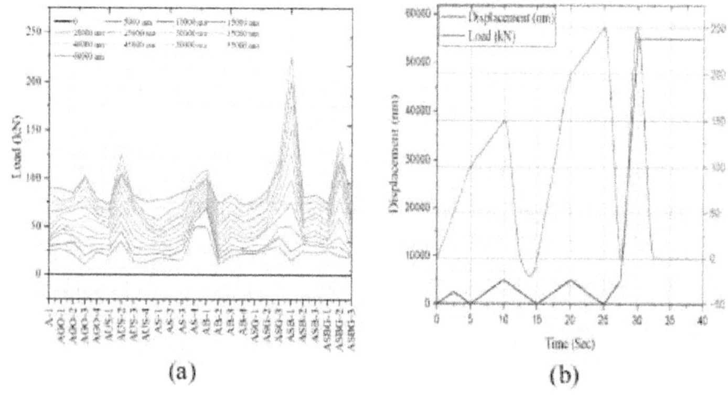

Figure 3.10. (a) Failure mode of selected samples and (b) Time taken for the best specimen.

3.4.5. Root bending stress

Due to the transmission load, root bending stress has occurred at the tooth's root in the gears. Bending stress is instigated on gear teeth due to tangential force generated by the energy. The bending stress is calculated by Equation 3.11. The following Equation estimates where F_t is a tangential force acting on the gear tooth (as represented in Equation 3.12).

$$\sigma_b = \frac{F_T}{b_1 m_n} q_e q_k \quad \text{.......................... } \textit{Equation (3.11)}$$

$$F_t = \frac{T_1}{\frac{1}{2} d_p} \quad \text{.......................... } \textit{Equation (3.12)}$$

Here, q_e is a factor from the contact ratio, which Equation 3.13 calculates. q_k is the tooth dorm factor, which the following Equation 3.14 calculates,

$$q_e = \frac{1}{CR} \quad \text{.......................... } \textit{Equation (3.13)}$$

$$q_k = \frac{6 m_n h_q \cos a'_a}{S_q^2 \cos a} \quad \text{.......................... } \textit{Equation (3.14)}$$

The permissible root bending stress for a material is indicated byσ_{bp}. The stress at the root should be less than the bending stress. The following Equation 3.15 calculates the minimum modulus,

$$m > \sqrt[3]{\frac{200 T_1 q_e q_k}{\frac{b_1}{d_{p,1}} z_1^2 \sigma_{bp}}} \quad \text{.......................... } \textit{Equation (3.15)}$$

Identical expression with the wheel specifications provides the wheel's root stress. Usually, root stress on the wheel is lower than the pinion. The gear tooth strength graph shows the hardness to tensile strength of the composition of our selected mix. The ASB-1 type shows a high gear tooth strength compared with all other materials, which is graphically described in Figure 3.11. The error rate of the tested specimens is illustrated in Figure 3.12, which shows that the ASB-1 sample has the lowest error rate among them.

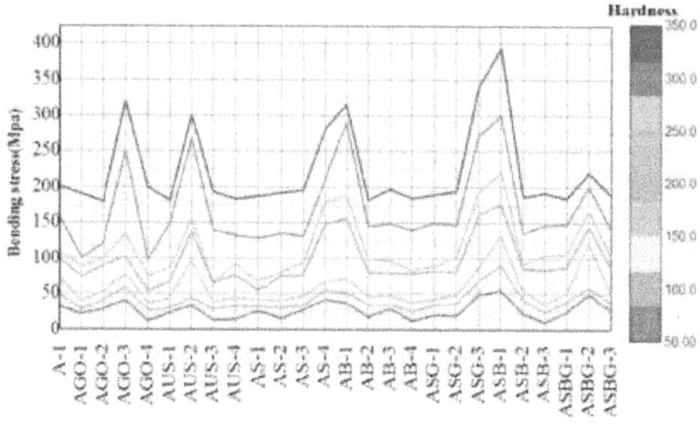

Figure 3.11. Gear tooth strength.

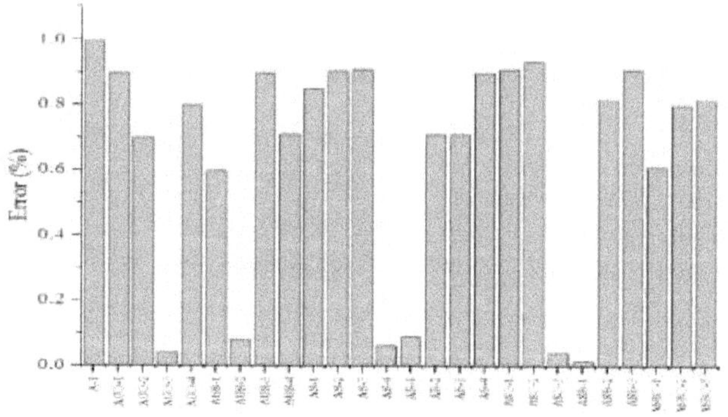

Figure 3.12. Error rate.

3.4.6. Thermal expansion

The coefficient of thermal expansion (CTE) is one of the significant properties of composite materials. The calculation of thermal expansion is necessary to know about the material's function. It is related to the material utilized at maximum and minimum temperature, as the size variation resulting from the thermal expansion. While the material is reinforced with other coating elements to slow down and decrease the diffusion of the material, the material is tested concerning temperature. CTE is calculated with different temperature values ranging from 100 to 1500°C. The CTE is evaluated using Equation 3.16.

$$CTE = \frac{C}{L_i}\left(\frac{\Delta L_j}{\Delta T_j}\right) \quad \text{........................... } \textit{Equation (3.16)}$$

Here, CTE is the coefficient of thermal expansion in μm/m, C is the consonant about 1, L_i is the initial length in m, ΔL_j is the change in length in μm, and ΔT_j is the temperature change. The CTE graph is represented in Figure 3.13. Here, the ASB-1 specimen attained lower thermal expansion than the samples. Hence, the ASB-1 specimen has high strength in thermal resistance even at high temperatures.

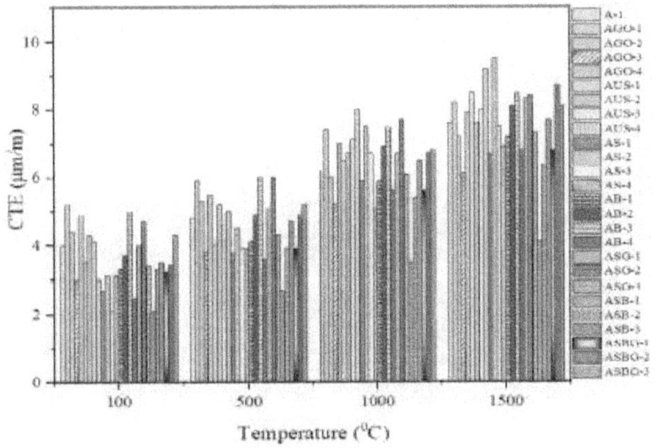

Figure 3.13. Assessment of CTE.

3.4.7. Surface roughness

Surface roughness is investigated to determine the irregularities in the surface during the machining process since aluminum composites are abrasive. Surface roughness values unveil the function of composite material after the machining process. The surface roughness was evaluated using the Mitutoyo company surtronic 3+ measuring tool. The least count of R*a is 1 micron, and the cut length is about 0.8 millimeters. Hence, the changes in surface roughness for all process functions, concentrations of reinforced ceramic particles, speed, feed level, and depth of cut are proved in Figure 3.14. The surface roughness is reduced with increased speed. The surface roughness is enhanced with enhancement in the feed level. The surface roughness increases with the depth of the cut;

as the concentration ratio of alumina particles increases, the surface roughness increases because of the abrasive function of alumina.

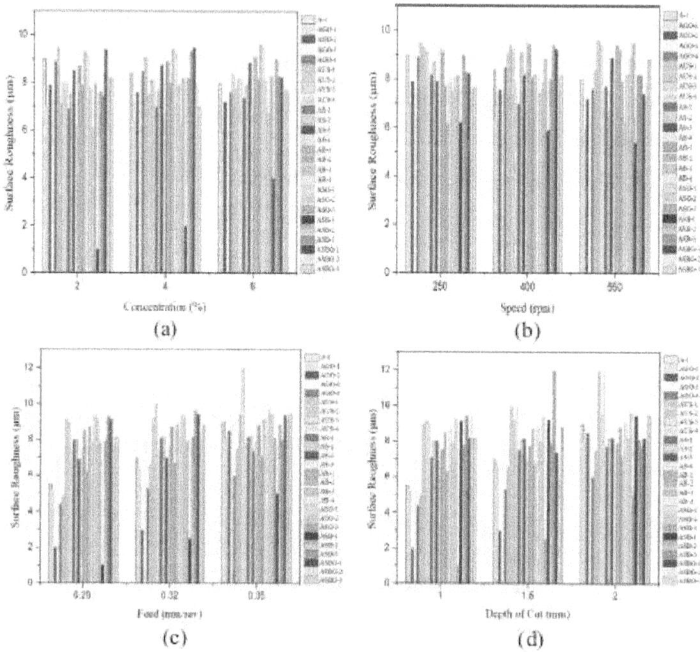

Figure 3.14. Surface roughness (a) with concentration, (b) with speed rate, (c) with feed level, and (d) with a depth of cut.

3.4.8. Comparison assessment

In addition, the reason for choosing Strawberry optimization over other optimization is defined in Figure 3.15. Various effective optimizations, such as Particle Swarm Optimization (PSO) [117], Strawberry Optimisation (SBO) [115], Genetic Algorithm (GA) [118], Bat Algorithm (BA) [118], Ant Colony Optimization (ACO) [119], and African Buffalo Optimization (ABO) [118] has been considered. The testing was done so that the strawberry solution recorded the finest outcome. Hence, the strawberry optimization model is taken for this study.

The impact strength and tensile strength of the adequate composite specimen are compared with other existing hybrid samples to determine the effectiveness of the hybrid aluminum matrix composite specimen. Moreover, the proper composite model is compared with other existing hybrid composites like Al-Mg-Si-SiC/muscovite/T6, Al-SiC, and Al-TiO_2-H, SiC-FA-Al and A-1 (pure-Al 6061). The comparison assessment is shown in Table 3.9.

The comparison result indicated that the Al-Mg-Si- SiC/muscovite/T6 had attained the impact and tensile strength of 10 J and 130 MPa, respectively, less than the prepared hybrid composite specimen. In addition, the Al-SiC and SiC-FA-Al have achieved tensile strength of 157 MPa and 184.57 MPa, respectively. Moreover, the Al-TiO_2-H has attained a tensile strength of 160.20 MPa and an impact strength of 34 J, less than the prepared ASB-1 specimen. The A-1 specimen obtained the tensile and impact strength as 216.82 MPa and 25.75 J, respectively. Consequently, the effective hybrid composite specimen ASB-1 has attained an impact strength of 36 J and tensile strength of 325.6 MPa, which are higher than all other specimens.

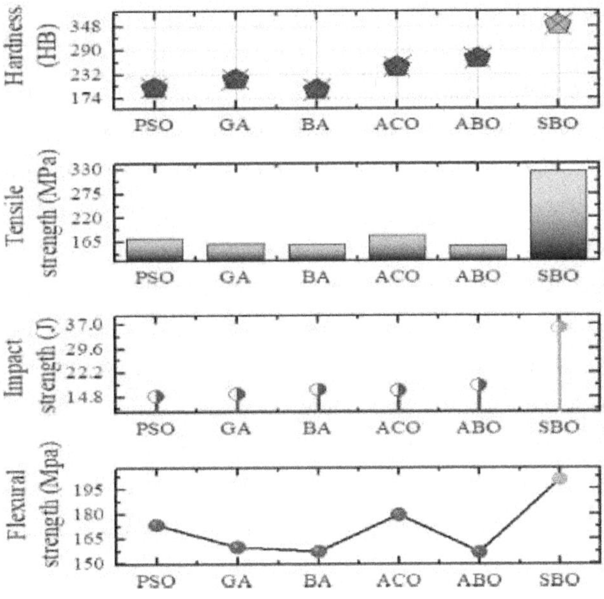

Figure 3.15. Effectiveness of optimization.

Table 3.9. Comparison assessment

S. No	Composites	Tensile strength (MPa)	Impact strength (J)
1	Al-Mg-Si-SiC/muscovite/T6	130	10
2	Al-SiC	157	-
3	Al-TiO_2-H	160.20	34
4	SiC-FA-Al	184.57	-
5	A-1	216.82	25.75
6	Proposed (ASB-1)	325.6	36

3.4.9. Discussion

Al 6061-SiC-B4C-GO hybrid nanocomposites exhibit excellent mechanical properties, including high strength and toughness, making them suitable for various applications, such as aerospace and automotive industries. Optimizing the mix proportion of these hybrid composites is critical for achieving the desired properties. The traditional mix selection approach is limited in handling complex problems and may not always produce the best mix proportion. To overcome these limitations, a novel approach that combines a strawberry optimization algorithm with a recurrent neural network can be used. In this approach, the strawberry optimization algorithm searches for the optimal mix proportion of the hybrid nanocomposites. The recurrent neural network is used to model the relationship between the mix proportion and the mechanical properties of the hybrid sites. The recurrent neural network is trained using experimental data, and the trained model is used to predict the mechanical properties of the hybrid sites for different mix proportions. The mechanical and tribological properties of the predicted ASB-1 with a strawberry optimization and recurrent neural-based work mix selection approach attained high mechanical strength, including its strength, stiffness, and toughness. The tribological properties of the ASB-1 composite material achieved lower tribological properties, including its wear resistance and coefficient of friction. To show the efficiency of the proposed mix selection, the materials taken for comparison and the proposed mix are tested in the same platform. The materials' managerial implications are found in terms of tensile strength in MPa, flexural strength in MPa, impact strength in J, hardness in HB, wear rate in mm^3/Nm, coefficient of friction, and abrasion, and the values are tabulated in Table 3.10. These values are evidence of the effectiveness of the proposed mixture. The proposed mix attains the highest tensile, impact, flexural, and hardness values, lower wear rate, and coefficient of friction and abrasion. The comparison values of the proposed composite in terms of other composites are mentioned in Table 3.11.

The test outcomes indicate that ASB-1 has achieved higher strengths, less abrasive friction and friction coefficient, and less wear rate. It has proved that the 86 wt.% Al 6061 + 2 wt.% GO +10 wt.% SiC +2 wt.% B4C (ASB-1) hybrid composite specimen has enhanced its strengths and is selected for gear manufacturing. This work uses different parameters to measure the prepared specimen's performance. The ASB-1 is chosen as an adequate man due to its improved strengths. Moreover, the proper hybrid ASB-1 composite sample is compared with other composite models. The overall performance of ASB-1 has attained higher tensile strength, flexural strength, impact strength, less abrasive friction coefficient, wear rate, and friction coefficient. Also, the gear tooth strength is calculated using the Lewis equation, and microstructural characteristics are computed using SEM. Hence, utilizing these materials wt.%, the aluminum matrix gear has been prepared. It has proved that the selected optimal mix is suitable for the gear application. The data set for model training should differ from the predicted output optimal compositions. This helps to ensure that the model is not simply memorizing the data and can accurately generalize to new compositions. The algorithm was trained on 26 compositions, and 26 compositions were experimentally synthesized. While this approach may have limitations regarding generalizability to new compositions, the study aimed not to generalize to new compositions but to optimize the properties of the specific Al 6061-SiC-B_4C-GO hybrid nanocomposites system under investigation. However, it is essential to note that the model's generalizability to new compositions is vital in materials science research. Therefore, future studies could aim to test it with a practical scenario to give the optimal solution's robustness and need. Additionally, the study could be expanded to include a larger data set for training the algorithm, which may improve the model's generalizability to new compositions.

Table 3.10. COF with varying load

Load	COF				AFC			
	10	**20**	**30**	**40**	**10**	**20**	**30**	**40**
A-1	0.72	0.76	0.82	0.9	0.71	0.74	0.79	0.83
AGO-1	0.81	0.83	0.88	0.93	0.73	0.75	0.79	0.845
AGO-2	0.82	0.86	0.9	0.923	0.75	0.79	0.802	0.831
AGO-3	0.8	0.82	0.84	0.86	0.71	0.73	0.76	0.79
AGO-4	0.86	0.9	0.91	0.941	0.79	0.81	0.85	0.88
AUS-1	0.9	0.92	0.95	0.982	0.8	0.82	0.84	0.87
AUS-2	0.71	0.73	0.77	0.8	0.61	0.63	0.67	0.7
AUS-3	0.85	0.88	0.902	0.919	0.81	0.84	0.86	0.878
AUS-4	0.907	0.93	0.956	0.972	0.83	0.87	0.89	0.91
AS-1	0.891	0.93	0.961	0.989	0.84	0.87	0.911	0.93
AS-2	0.91	0.93	0.952	0.961	0.903	0.925	0.956	0.971
AS-3	0.88	0.89	0.91	0.934	0.72	0.76	0.81	0.846
AS-4	0.71	0.73	0.76	0.79	0.69	0.71	0.73	0.75
AB-1	0.71	0.73	0.75	0.77	0.51	0.53	0.58	0.61
AB-2	0.9	0.92	0.941	0.958	0.81	0.84	0.86	0.875
AB-3	0.91	0.93	0.947	0.965	0.804	0.83	0.851	0.889
AB-4	0.934	0.958	0.972	0.993	0.751	0.783	0.815	0.839
ASG-1	0.82	0.85	0.87	0.915	0.81	0.83	0.85	0.863

ASG-2	0.911	0.922	0.941	0.951	0.808	0.826	0.871	0.891
ASG-3	0.61	0.65	0.68	0.72	0.46	0.48	0.51	0.55
ASB-1	0.402	0.421	0.442	0.46	0.401	0.422	0.462	0.49
ASB-2	0.91	0.94	0.96	0.976	0.9	0.91	0.93	0.945
ASB-3	0.92	0.94	0.96	0.989	0.83	0.85	0.88	0.923
ASBG-1	0.924	0.934	0.952	0.98	0.903	0.913	0.934	0.94
ASBG-2	0.61	0.63	0.67	0.7	0.51	0.53	0.58	0.6
ASBG-3	0.91	0.94	0.97	0.99	0.89	0.91	0.93	0.95

Table 3.11. Performance evaluation

S. No	**Composites Test**	**Al-Mg-Si-SiC/muscovite/T6**	**Al-SiC**	**Al-TiO_2-H**	**SiC-FA-Al**	**A-1**	**Proposed (ASB-1)**
1	Tensile strength (MPa)	130	157	160.20	184.57	216.82	325.6
2	Impact strength (J)	10	28	34	33	25.75	36
3	Flexural strength (MPa)	150.89	167.45	170.23	178.34	188.72	200.75
4	Hardness (HB)	245.22	239.56	275.43	292.22	300.22	350.45
5	Wear rate (mm^3/Nm)	0.01	0.04	0.1	0.08	0.2	5.54×10^{-3}

6	Friction coefficient	0.58	0.62	0.69	0.71	0.5	0.46
7	Abrasive friction coefficient	0.60	0.65	0.70	0.73	0.52	0.49

3.5. Conclusion

This research has introduced and applied an innovative approach called Strawberry Recurrent Neural-based Mix Selection (SRNMS) to determine the optimal mix proportions for aluminum composite materials. The key outcomes of these investigations are summarised as follows:

Selection of mix proportions: Various combinations of B_4C, S_iC, and GO reinforcements in Al 6061 aluminum matrix composites were considered. The SRNMS method was employed to identify the most suitable mix proportions for fabricating specimens.

Surface coating with TiO_2: To prevent aggregation and enhance bonding strength, a coating of TiO_2 was applied to SiC using the EPD method. The identified optimal mix proportions were successfully used in the stir-casting process. Furthermore, the study assessed the tribological and mechanical properties of the composite specimens.

Identification of adequate specimen: The results indicated that ASB-1 is a suitable specimen for manufacturing gears, highlighting the effectiveness of the developed SRNMS framework in identifying mix proportions and potentially reducing the overall production costs.

SEM analysis: SEM examination demonstrated a uniform distribution of reinforcements within the aluminum matrix. This uniform distribution was attributed to preheating the reinforcements before pouring, ensuring good interfacial bonding, and employing an appropriate stirring speed.

Worn surface analysis: The worn surface analysis revealed that ASB-1 exhibited a smoother surface due to the crushing of reinforced particulates, work-hardened pin surfaces, and the formation of a protective layer between the counter-plate and pin surface. Furthermore, the optimization-based reinforcement components proposed in this study yielded favorable results in composite tests, surpassing other mixing properties. Notably, the composite achieved a tensile strength of 325.6 MPa, an impact strength of 36 J, and a flexural strength of 200.75 MPa. Additionally, coefficient tests validated the effectiveness of the ASB mixing, showing a 1.04% decrement in both friction and abrasive friction coefficients. Moreover, the ASB mixing demonstrated a hardness of 350.45 HB and a wear rate of 5.54×10^{-3} mm^3/Nm.

In conclusion, using SRNMS for mix proportion determination and the subsequent fabrication of composite specimens with enhanced properties have showcased the potential for cost-effective and efficient composite manufacturing processes. The results suggest that the developed approach can significantly impact industries reliant on advanced composite materials, particularly in applications where wear resistance and mechanical performance are critical.

CHAPTER-4

4. PERFORMANCE EVALUATION OF COMPOSITE GEARS

The present objective offers a comprehensive method for assessing the contact stress in a hybrid nano-composites gear made of Al 6061-SiC-B_4C-GO using experimental testing and Finite Element Analysis (FEA). The experimental testing was carried out on a specialized test rig that can apply different misalignments to the gears and use strain gauges to measure the contact stress. The contact stress distribution on the gear teeth is predicted using the FEA simulations. There is an acceptable level of agreement, proving the effectiveness and accuracy of the integrated approach when the findings from the two techniques were compared. The study shows the efficacy of an integrated strategy for assessing the contact stress of gears under misalignment. It offers valuable insights into the behavior of contact stress in aluminum composite gears. The findings of this study can be used to improve the design and performance of aluminum composite gears, leading to more efficiency and reliability even under misalignment for machinery.

4.1.Fabrication of spur gears

The fabrication of aluminum composite spur gears by Wire EDM started with selecting the optimal material mix, i.e., ABS-1, consisting of 86 wt.% Al 6061, 2 wt.% GO, 10 wt.% SiC, and 2 wt.% B_4C. This composite mix has been chosen for its excellent properties, including enhanced tensile, impact, and flexural strengths, along with a low friction coefficient and reduced wear. The established properties of ABS-1 composite are displayed in Tabel 4.1. As stated earlier, the materials were mixed and subjected to a stir-casting process to achieve a homogenous distribution of the reinforcements. After casting, the composite was shaped into gear blanks. These blanks were then precisely machined using Wire EDM at Maruthi Tooling, where the gears were cut to the required specifications with high accuracy and minimal thermal distortion, ensuring optimal performance in mechanical applications. The specification of the composite gear are documented in Table 4.2. The photographs of the prepared composite gears are shown in Figure 4.1.

Table 4.1. Established Properties of ABS-1 composite

Parameters	Obtained outcome
Tensile strength	325.6 MPa
Impact strength	36 J
Flexural strength	200.75 MPa
Wear rate	5.69 10^{-3} mm^3/Nm
Friction coefficient	0.51

Table 4.2. Specification of composite spur gear

Parameters	Gear/pinion
Module (mm)	5
Face width (mm)	20
Pressure angle (Degree)	20
No. of tooth	20
Pitch circle diameter (mm)	100
Fillet radius (mm)	1.25
The outer diameter of the gear (mm)	110
Addendum (mm)	5
Dedendum (mm)	6.25
Internal shaft diameter (mm)	20

Figure 4.1. Prepared composite gears

4.2. Test rig

The experimental test rig was designed and manufactured to measure the contact stress of the gear under various loading conditions. The test rig typically consists of a gear train that simulates a real-world application and applies a specified load to the gears. The gear test rig has sensors and measurement tools that allow observing and analyzing the gears' behavior under various loads and operating conditions. The data collected from the gear test rig can be used to determine the gear's strength, durability, noise, vibration, and efficiency. The results of the gear test rig can be used to optimize the design of the gear and improve its performance in real-world applications. The photograph of the test rig is displayed in Figure 4.2. The test rig was made to have a provision to study three types of misalignments, such as angular, radial, and axial, along with their combinations.

Figure 4.2. Photograph of experimental gear test rig

The test procedure for the gear test rig can vary depending on the specific application and the type of gear being tested. In the current work, the gearbox is operated at a speed of 100 rpm with different torques between 45 and 100 Nm at regular intervals. A significant advantage of the gear test rig is that the alignment of the carbon slip ring allows strain gauge readings to be read on real-time contact teeth during the gear rotation time. Figure 4.3 shows the arrangement of the gear.

Figure 4.3. Arrangement of gears

The carbon slip ring is an electrical component that transfers power and data signals between a rotating and a stationary part in a machine or device. The carbon slip ring is designed to provide a continuous and reliable electrical connection while the rotating part of the machine is in motion. A carbon slip ring typically consists of a rotating ring and stationary brushes that contact the ring to transfer electrical power and signals. Strain gauge connections from rotating system parts, such as transmission shafts, are connected to the ring. The strain gauge arrangement is shown in Figure 4.4. The present study aims to determine the contact stresses between two mating gears, which are evaluated by measuring the corresponding strain of the gears. The strain measurement can be done using various methods. The usual method is to utilize resistance strain gauges. The resistance strain gauge works on the principle of change in resistance when subjected to a load. These strain gauge sensors are highly precise and give values with minimum division. The accuracy of the strain gauge is $\pm 2\%$.

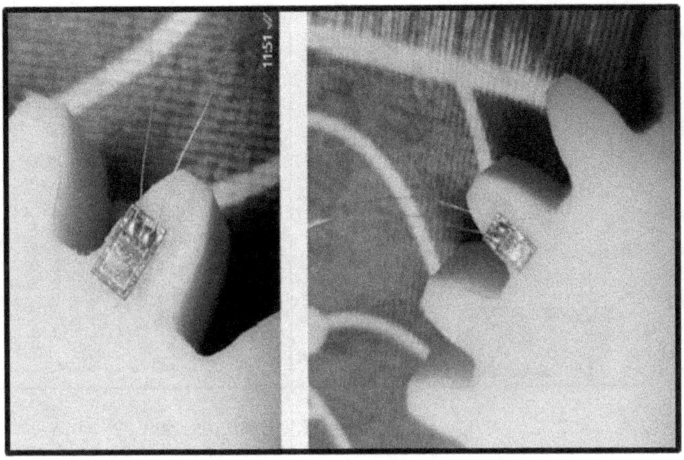

Figure 4.4. Arrangement of strain gauge

The strain gauge sensors are bonded to the gear teeth surface using specific additives. Measurement of the change in resistance requires an elaborate circuit, as shown in Figure 4.5. The circuit consists of an SMPS to reduce the voltage from 230 to 10 V and rectify it from AC to DC. The DC voltage from SMPS is transmitted through one set of slip rings to the quarter bridge strain gauge circuit as excitation voltage. The output voltage generated from the unbalanced bridge is amplified by a circuit fixed to the rotating shaft. A second set of slip rings is used to transmit the amplified output voltage to the Arduino Uno. The Arduino Uno converts the analog output voltage to a digital value.

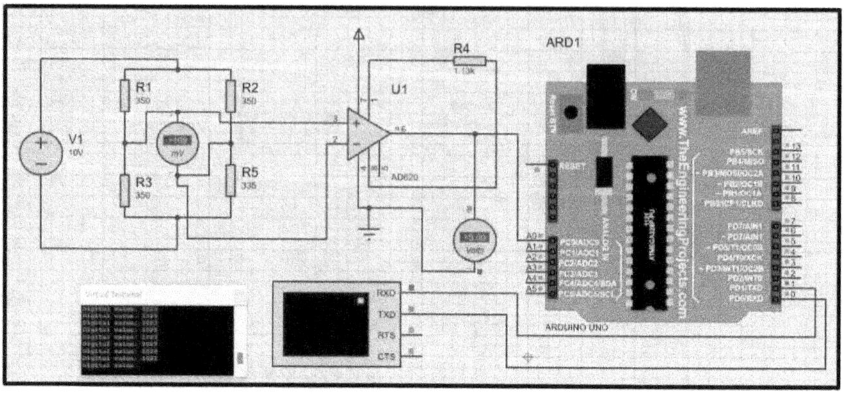

Figure 4.5. Circuit diagram

4.3. Calculation of stress

The strain gauge sensor in the circuit with a normal resistance value of 350 Ω was used. The gauge factor (GF), which is the fundamental parameter to assess its sensitivity, is quantitatively expressed as the ratio of fractional change in electrical resistance to the fractional change in length (i.e., strain) given the following equations;

$$\mathrm{GF} = \frac{\Delta \mathrm{R}}{\mathrm{R}} / \frac{\Delta \mathrm{L}}{\mathrm{L}} \quad \text{-----------} \quad (\textit{Equation-4.1})$$

$$GF = \frac{\Delta R}{R} / \varepsilon \quad \text{--------------} \quad (\textit{Equation-4.2})$$

$$\varepsilon = \frac{\Delta \mathrm{R}}{\mathrm{R}} / \mathrm{GF} \quad \text{----------------} \quad (\textit{Equation-4.3})$$

This expression relates the strain to the change in resistance. The gauge factor (GF) for metallic strain gauge is typically around 2. The gauge factor of the strain gauge in the current experimental setup has a value of 2.11.

The measurement of strain is achieved using a Wheatstone bridge circuit (as shown in Figure 4.6), and the measurement of strains of very small order requires the accurate measurement of very small changes in resistance. Such small changes in resistance produced in strain gauges can be best measured using a bridge configuration with a voltage or current excitation source. The Wheatstone bridge consists of four arms, three of which are equal resistances, and the fourth is the strain gauge's resistance. The bridge circuit is given an excitation voltage (V_{ex}) of 10V applied across the bridge. The output voltage Vo of the bridge is expressed as;

$$Vo = \left(\frac{R3}{R3+R4} - \frac{R2}{R1+R2}\right) \text{ x } V_{ex} \quad \text{-------------} \quad (\textit{Equation-4.4})$$

The output voltage will be zero when $\frac{R1}{R2} = \frac{R3}{R4}$ under this condition, the bridge circuit is balanced. As stated above, the R4 is replaced by an active strain gauge; any change in the strain gauge resistance unbalances the bridge and produces a non-zero output voltage. The above equation can be modified as follows;

$$Vo = \frac{-GF*\varepsilon}{4}\left(\frac{1}{1+GF*\varepsilon/2}\right) \text{ x } V_{ex} \quad \text{---------} \quad (\textit{Equation-4.5})$$

The output voltage is very minute in microvolts and must be amplified. The amplification of the voltage is obtained through a supplementary amplifier circuit. The amplified output voltage is transformed into a digital value using a micro controller board titled ARDUINO NANO. In Equation (3.5), V_O is the function of the strain; if V_O is known, the strain can be calculated, and this strain can be converted into stress using Hook's law [120].

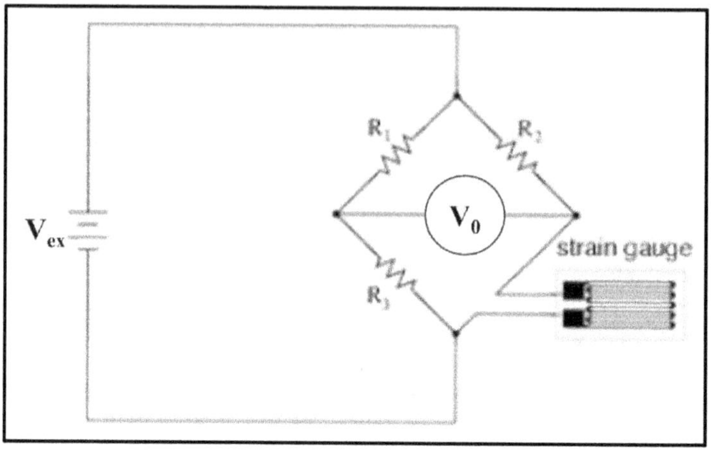

Figure 4.6. Wheatstone bridge circuit [120, 121].

4.4. 3D solid modelling

A 3D solid model of part of the aluminum composite gear, including the teeth and some parts of the gear body, was modeled using solid works. During the assembly of gears, misalignments such as radial, axial, and angular misalignments were provided in the modelling. Radial misalignment, the distance between the axis of the driving gear and the axis of the driven gear, is one-factor affecting gear performance. Figure 4.7a shows the gear assembly under radial misalignment of 3 mm. A radial misalignment of 3 mm is relatively small but must be addressed. Angular misalignment occurs when the axes of the driving and driven gears are not parallel and intersect at an angle. Figure 4.7b displays the gear assembly under angular misalignment of 1.5°. Axial misalignment occurs when the axis of the driving gear and the axis of the driven gear are not parallel. It can cause uneven wear, increase noise and vibration, and reduce gear life. Figure 4.7c represents the gear assembly under axial misalignment of 1.5 mm.

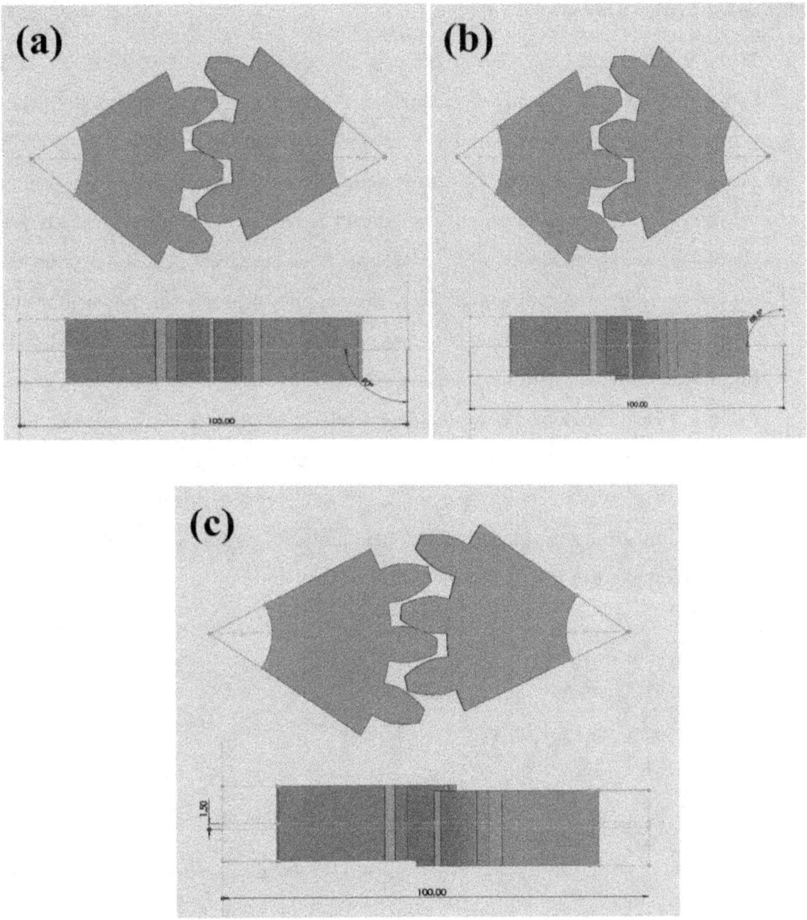

Figure 4.7. Gear Assembly under (a) Radial Misalignment of 3 mm, (b) Angular Misalignment of 1.5°, and (c) Axial Misalignment of 1.5 mm

4.5. Finite element analysis

In the present study, the FEA model was imported from solid works using ANSYS 14.5 software, and it is employed to analyze the gear under misalignments. The material properties, such as elastic modulus, Poisson's ratio, and yield strength, were defined based on the material properties of the gear material. Contact algorithms and constraints are used to simulate real-world scenarios. The contact algorithm used in this analysis was friction contact type, and constraints were fixed support for the driven gear and friction less support for the drive gear. A moment of 71 N-m is applied to represent power transmission. In ANSYS, meshing is dividing a complex geometry into smaller, more superficial elements for analysis. Mesh sensitivity analysis was done by performing many simulation trials so that the type and size of the mesh element selected would give accurate results; the mesh size of 2 mm and tetrahedron element type was selected for gear body geometry, and mesh size of 0.2 mm was selected at the contact of gears teeth to develop high mesh density, for the best simulation and accuracy. The total number of nodes used in this analysis is 29100, and the elements are 16770. Figure 4.8 a&b shows the constraints used and meshing during ANSYS.

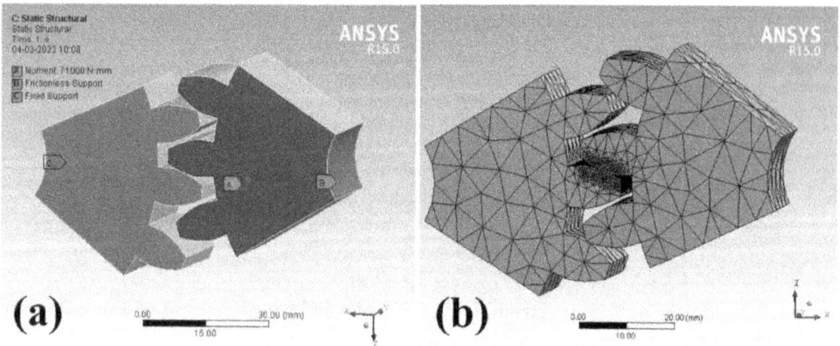

Figure 4.8. (a) Constraints used and (b) Meshing during ANSYS.

In summary, the methodology used in this study is an integrated approach that combines FEA and experimental test rig testing to analyze the contact stress behavior of aluminum composite gears. The introduced architecture is illustrated in Figure 4.9. The FEA model was used to obtain the contact stress distribution, while the experimental test rig measured the contact stress under different misalignment conditions. The gear performance was studied by comparing the FEA results with the experimental test rig findings.

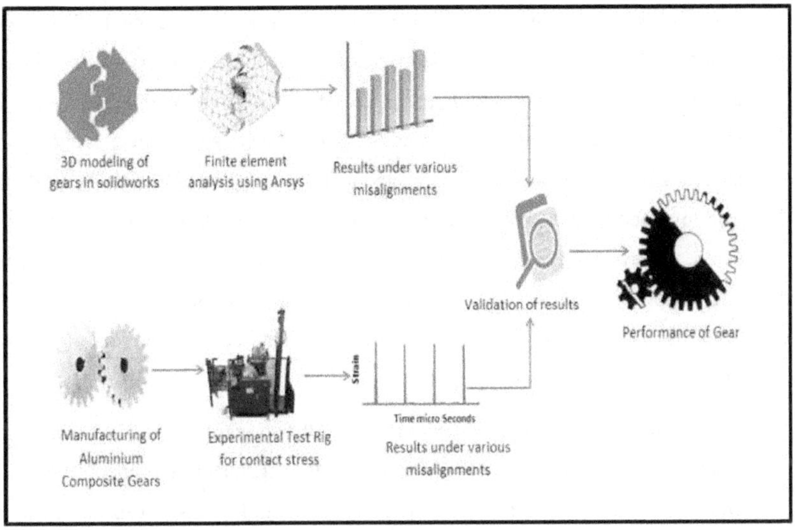

Figure 4.9. Proposed Architecture

4.6. Result and discussion

4.6.1. Contact strain

In order to represent the calculated strain value, the system was adjusted and configured. The system aims to maintain the strain gauge reading constant at the calibrated strain value until stability is reached. Any calibration mistakes were recorded and fixed as necessary when analyzing the other test results. The test was then conducted at 100 rpm. Figure 4.10 (a-d) represents the strain under perfectly aligned conditions, radial, angular, and axial misalignment of 3 mm, 1.5^{0}, and 1.5 mm, respectively. The average value was calculated by averaging ten peak readings. The corresponding strain values were assessed after considering any necessary error adjustments. The experimental contact strain is obtained by reading strain values from strain gauges positioned along the parting line's contact surface. Using the generalized Hooke's law and assuming plane stress in the contact area, the contact stress between gear tooth contacts was calculated using these experimental contact stress values. Contact strain measures the deformation of two mating surfaces in contact. It is typically expressed as the ratio of the actual area to the apparent contact area.

Under radial misalignment, the contact strain between the mating surfaces can increase significantly. This is because the contact area between the two surfaces is reduced, and the load is distributed over a smaller area. As a result, the stress on the contact area increases, leading to higher contact strain. The estimated contact strain under radial misalignment using analysis and test rig is documented in Table 4.3. Therefore, radial misalignment can lead to an increase in contact strain between the mating surfaces. To minimize the contact strain, it is essential to ensure proper alignment of the mating components and reduce any sources of misalignment as much as possible. In Figure 4.11, Radial misalignment Vs. Strain is diagrammatically represented, showing that the strain increases with an increase in the radial misalignment.

Angular misalignment occurs when the axes of two mating components are not parallel, resulting in a tilt or angle between them. Under angular misalignment, the contact strain between the mating surfaces can increase significantly. This is because the contact area between the two surfaces becomes a line contact instead of a surface contact. This results in a higher stress concentration on the contact line, leading to higher contact strain. The estimated contact strain under angular misalignment using analysis and test rig is documented in Table 4.4. Figure 4.12 shows the relation between the angular misalignment and strain.

Axial misalignment occurs when the axes of two mating components are not aligned, resulting in a gap or overlap between them in the axial direction. Under axial misalignment, the contact strain between the mating surfaces can increase significantly. This is because the contact area between the two surfaces becomes smaller and may even disappear in some areas. As a result, the load is distributed over a smaller area, leading to higher contact stresses and strain. The estimated contact strain under axial misalignment using analysis and test rig is documented in Table 4.5. As shown in Figure 4.13, The strain increases with an increase in the axial misalignment. To minimize the contact strain under axial misalignment, it is essential to ensure proper alignment of the mating components and reduce any sources of misalignment as much as possible.

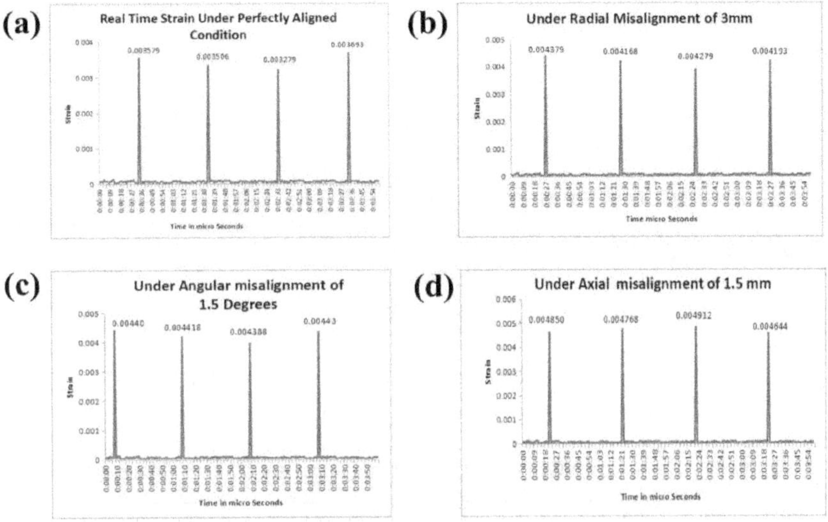

Figure 4.10. Strain under (a) perfectly aligned conditions, (b) Radial misalignment of 3mm, (c) Angular misalignment of 1.5° (d) Axial misalignment of 1.5mm

Table 4.3. Contact strain under radial misalignment

Radial Misalignment in (mm)	Strain From Analysis	Strain From Experimental Test Rig
0	0.003349	0.003514
1	0.003503	0.00368585
2	0.003745	0.003952
3	0.004058	0.004271

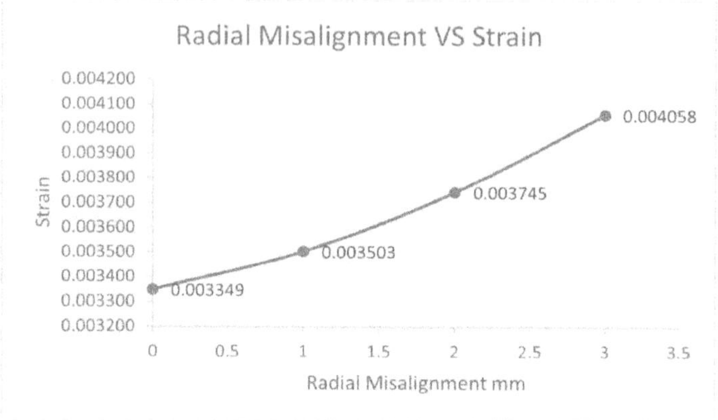

Figure 4.11. Radial misalignment Vs. strain

Table 4.4. Contact strain under angular misalignment

Angular misalignment in Degrees	Strain from analysis	Strain from the experimental test rig
0	0.003349	0.003514
0.5	0.003605	0.003782
1	0.003953	0.004058
1.5	0.004354	0.004413

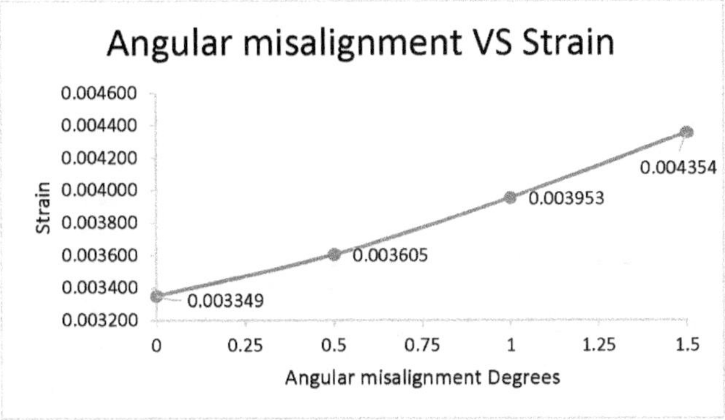

Figure 4.12. Angular misalignment Vs. strain

Table 4.5. Contact strain under axial misalignment

Axial misalignment mm	**Strain from analysis**	**Strain from the experimental test rig**
0	0.003349	0.003514
0.5	0.003679	0.003831
1	0.004035	0.004242
1.5	0.004440	0.004775

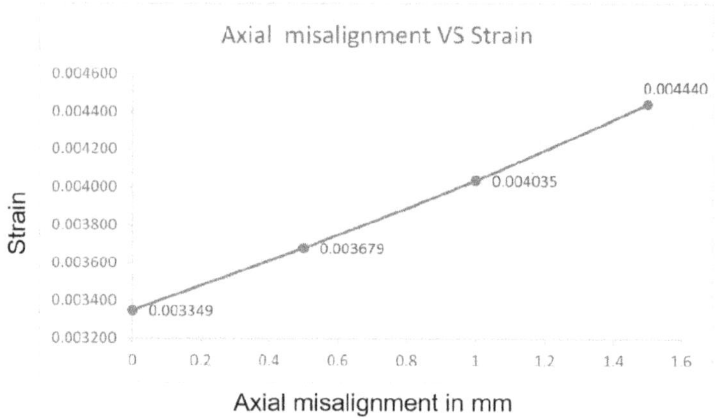

Figure 4.13. Axial misalignment Vs. strain

4.6.2. ANSYS

Figure 14 (a) indicates the ANSYS result under perfectly aligned conditions. When the ANSYS results are obtained under perfectly aligned conditions, in those cases, the simulation assumes that all model components are perfectly aligned, and there are no errors or misalignments. Under these conditions, the simulation results reflect the ideal behavior of the system being modeled. Figure 14 (b-d) indicates the ANSYS results under radial, angular, and axial misalignment of 3 mm, $1.5^{0,}$ and 1.5 mm. Like ideal conditions, it also means that the simulation assumes a radial, angular, and axial misalignment of 3 mm, $1.5^{0,}$ and 1.5 mm between the two gears being modeled.

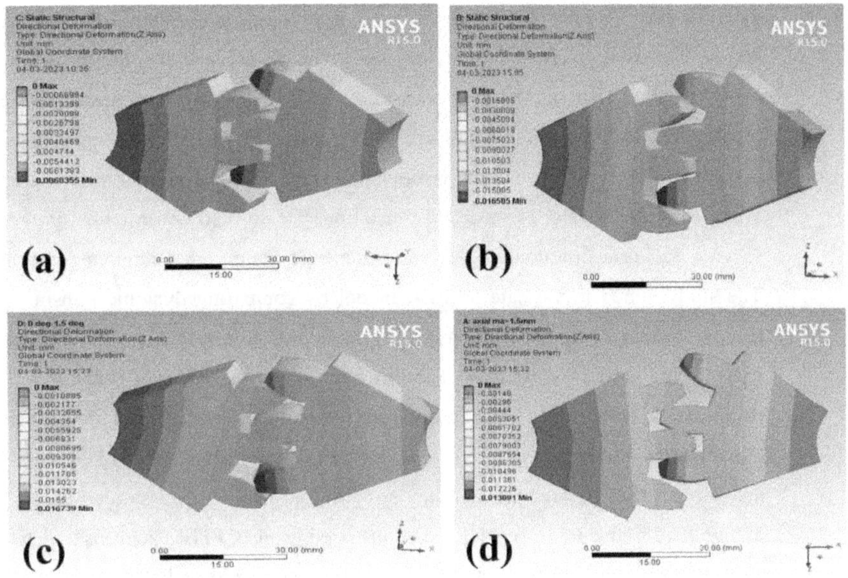

Figure 4.14. (a) Ansys result under perfectly aligned condition, (b) under radial misalignment of 3 mm, (c) under angular misalignment of 1.5° & (d) axial misalignment of 1.5mm.

The current research findings were compared with the existing literature studies. Qin et al. [10] investigated surface and subsurface stresses of gears subjected to cyclic shear and compressive stresses, leading to contact fatigue, a significant cause of gear tooth flank failure. Hertzian concept and finite elemental analysis (FEA) have been employed to forecast the occurrences of load cycles needed for fatigue fracture initiation depending on the multi-axial endurance mode. The investigation also explored the influence of the coefficient of friction (COF) and velocity on cyclic stresses and endurance limit, explaining that COF shifts the Von-mises load distributions, alters the stresses under shear loading, and reduces endurance limit. Finally, the researchers concluded that the loads close to the contacting and recess regions are higher than static contacting circumstances, leading to a decrease in endurance limit, mainly at higher velocities. FEA was employed by Chen et al. [26] to execute a loaded gear contact examination of a changed curvilinear gear set having localized bearing contact. The primary goal of this study is to examine the contact stresses and transmission errors under loading conditions. To achieve this, the study developed a mesh generation program based on a numerical model of a curvilinear gear produced employing a male fly cutter. Finally, the study presented numerical examples to explain the contact stresses and transmission errors underneath several designing variables and loads. Wang et al. [4] developed theoretical equations for no-load transmission error and time-varying backlash in a double-gear eccentricities system with a random contacting ratio. The study aimed to validate an FEA model by comparing dynamic transmission error results without gear eccentricity and predicting dynamic transmission error with eccentricities and varying loads. The study examined the effect of backlash on dynamic transmission error under 3 conditions using varying sinusoidal loads on gear pairs. The dynamic transmission error curve behavior was determined as the load direction changed and compared to no-load transmission error and no-load transmission error plus backlash at high load frequency. Li et al. [3] demonstrated using 3D FEM analysis to determine the surface contacting and root bending stresses for a pair of parallel-shaft spur gears. They introduced machining errors, assembly errors, and tooth modifications into a 3D coordinate system. They used a face-contact model to assume tooth contact on a reference face around the geometrical contacting line. By conducting loaded tooth contact analysis using FEA programs, they calculated deformation influence coefficients, determined tooth contract lengths, and analyzed tooth contacting patterns and root strains. The results were compared with the measured ones, demonstrating that machining errors, assembly errors, and tooth modifications significantly affect surface contact and root bending stresses. The present study aligns with the above-mentioned results and yielded better results. Validation of data is a critical step in this present research. The accuracy of the finite element analysis simulations is validated by comparing the results obtained with experimental measurements conducted at different levels of different misalignments. The percentage of deviation of results is significantly less, and the value is approximately less than 10%, which ensures that the FEA data is more accurate and reliable.

4.6.3. Summary

The results of the integrated approach for contact stress analysis of aluminum composite gears under misalignment showed that the contact stress behavior of the gears is significantly affected by misalignment, and the following conclusions have been extracted.

Comparing FEA and experimental results, it was observed that the FEA model could predict the contact stress behavior of the gears under misalignment. The experimental test rig results showed that the contact stress increased with increasing misalignment, which is consistent with the FEA results. The FEA model was able to predict the contact stress distribution on the tooth surface, which matched well with the experimental test rig results. Several experiments were also conducted to examine the contact stress behavior of aluminum composite gears under various misalignments. It was shown in this investigation that contact stress increased as the misalignment angle increased. When the misalignment value was higher, it affected the contact stress behavior more.

Through a better comprehension of the contact stress behavior, the results of this study can be used to optimize the design and material choice of aluminum composite gears under misalignment. According to the study, misalignment should be considered while designing and analyzing aluminum composite gears, and the proper steps should be taken to minimize misalignment.

FEA and experimental test rig testing were combined to successfully analyze the contact stress behavior of aluminum composite gears under misalignment. The investigation results demonstrated that gear misalignment significantly impacts how they react to contact stress. It is essential to take the proper precautions to minimize misalignment in the construction and operation of such gears. The study findings offer insightful information about the design and choice of materials for aluminum composite gears under various misalignments.

CHAPTER-5

5. CONCLUSIONS AND FUTURE SCOPE

5.1.Conclusions

- The present investigation introduced the Strawberry Recurrent Neural-based Mix Selection (SRNMS) approach to optimize mix proportions for aluminum composite materials, and it selected optimal mix proportions of B_4C, SiC, and GO reinforcements in Al 6061 composites using SRNMS, which were successfully applied in stir-casting.
- The investigation enhanced bonding strength by applying a TiO_2 coating to SiC via the EPD method, improving the composite's tribological and mechanical characteristics, and ASB-1 was identified as the most suitable specimen for gear manufacturing, demonstrating the cost-reduction potential of the SRNMS framework.
- Homogeneous reinforcement distribution in the aluminum matrix was obtained through preheating and appropriate stirring, as confirmed by SEM analysis and worn surface analysis revealed that ASB-1's smoother surface, high tensile strength, impact strength, flexural strength, and hardness resulted from optimized reinforcement and processing.
- The study demonstrated a 1.04% decrement in friction coefficients and a wear rate of 5.54×10^{-3} mm^3/Nm, highlighting the effectiveness of the ASB mix. It also demonstrated that SRNMS significantly impacts cost-effective and efficient composite manufacturing, particularly in applications requiring enhanced wear resistance and mechanical performance.
- The contact stress analysis of aluminum composite gears under misalignment confirmed that contact stress behavior increases with the misalignment angle. The FEA model's ability to predict contact stress distribution was validated and aligned well with the experimental test rig results. The study emphasized the need to consider misalignment in gear design and material selection, offering insights for minimizing its effects on construction and operation.

5.2. Future scope

- Applying advanced coatings to gear tooth surfaces to study their impact on reducing contact stresses under misalignment, enhancing gear performance and durability.
- Examining the impact of teeth modification on various Aluminium gear materials under misalignment conditions.
- Augmenting the gear test rig setup with additional accessories for measuring vibrations, noise levels, and temperature induced at the contact surface.

www.ingramcontent.com/pod-product-compliance
Lightning Source LLC
LaVergne TN
LVHW021737071126
851857LV00040B/1472
9798240816697